Methods in Animal Behaviour

Marc Naguib · Gretchen F. Wagner · Lysanne Snijders ·
E. Tobias Krause

Methods in Animal Behaviour

Marc Naguib
Behavioural Ecology Group
Wageningen University & Research
Wageningen, Gelderland, The Netherlands

Gretchen F. Wagner
Behavioural Ecology Group
Wageningen University & Research
Wageningen, Gelderland, The Netherlands

Lysanne Snijders
Behavioural Ecology Group
Wageningen University & Research
Wageningen, Gelderland, The Netherlands

E. Tobias Krause
Institute of Animal Welfare and Animal
Husbandry
Friedrich-Loeffler-Institut
Celle, Niedersachsen, Germany

ISBN 978-3-662-67791-9   ISBN 978-3-662-67792-6   (eBook)
https://doi.org/10.1007/978-3-662-67792-6

Foreword

We thank Chris Tyson for helping us substantially with ▶ Sect. 6.3. Furthermore, are we thankful to Oliver Krüger, Antonia Patt and Christopher Schutz for providing wonderful photos for this book.

Marc Naguib
Gretchen F. Wagner
Lysanne Snijders
E. Tobias Krause
Wageningen and Celle
May 2023

Contents

Introduction

Contents

1

1.1 What Animal Behaviour is and Why it is Important

Interest in the behaviour of animals is probably as old as humanity itself, as animals are an integral and omnipresent part of our natural environment. For centuries, an understanding of the behaviour of animals has played a central role in hunting, and particularly through the domestication of companion and farm animals, the behaviour of some species has even evolved to be intertwined with our own (Darwin 1859; Nicol 2015). Interest in and knowledge of animal behaviour are therefore closely linked to the evolutionary development of humankind. Particularly since Darwin and Wallace (Browne 2013) and the basic understanding of evolutionary processes, observing and studying animal behaviour has gained importance as a scientific discipline. Due to the diversity of dimensions at which animal behaviour can be studied and understood, behavioural research has progressed within the past century into a lively and rapidly growing field. One of the main reasons for this is the steady development of sophisticated quantitative methods that are used in the study of animal behaviour.

Modern animal behaviour research is a highly quantitative scientific discipline in which a wide range of specialized, often experimental methods for recording and assessing animals' behaviours has been developed. Consequently, sophisticated data analysis and statistical skills have become part of the essential spectrum of methods in this field. Moreover, the enormous diversity of animal behaviour research stems from, among other things, the fact that within this discipline both the immediate proximate mechanisms underlying behaviour and the function and evolution of behaviour are of interest. As the behaviour of an organism is determined and influenced by a variety of internal processes (e.g. genetic, physiological, endocrinological and neurobiological) as well as external factors (e.g. ecological and social variables), the range of methods used in animal behaviour research spans from molecular biological, physiological, and neurobiological methods to ecological methods that shed light on the function and evolution of behaviour (◘ Fig. 1.1).

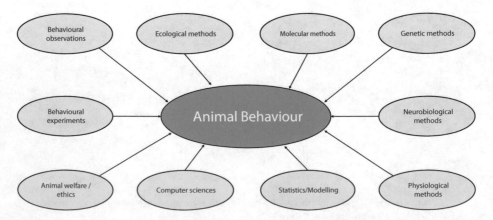

◘ **Fig. 1.1**   Scheme of other research areas from which methods are applied in animal behaviour research

With such a wide scope of questions and corresponding methods from other research areas, one might wonder what holds animal behaviour research together as a scientific discipline. A central and unifying concept is research on the "whole" animal or to simulate them using modelling approaches (e.g., Krause et al. 2011; Chouinard-Thuly et al. 2017; Bierbach et al. 2018). Regardless of the methodological focus of a specific behavioural study, the aim is to understand behaviour at the level of the individual, whole animal.

The study of animal behaviour contributes to the understanding of biological processes, such as the development and selection of traits, as well as the functioning of an organism and, last but not least, our own behaviour (e.g., Danchin et al. 2008; Davies et al. 2012; Alcock 2013; Dugatkin 2019). The results of animal behaviour research have also significantly influenced social discourse. The early work was an important trigger for the debate about innate and acquired parts (nature *versus* nurture) of animal (including human) behaviour. In particular, sociobiological interpretations of human behaviour (Wilson 1975; Krebs and Davies 1997) and books such as the *Selfish Gene* by Richard Dawkins (1976) have triggered heated debates about our understanding of the world we live in and how we should see ourselves in it. In its increasingly close connection with evolutionary biology and ecology, as well as animal physiology and neurobiology, animal behaviour research had a major impact on how we view modern humans and how we understand the evolution and development of our behaviour. Much of what was regarded in the past to be uniquely human has since been found in our primate relatives (Byrne and White 1988; Tomasello and Call 1997), and also in many other animal taxa, such as birds (Emery and Clayton 2004; Emery 2016). The fields of sociobiology and behavioural ecology, in their integration of behaviour and evolution, have led to new insights into the adaptation of behaviour to environmental conditions and the often astonishingly rapid evolution of behaviour during environmental changes. Animal behaviour also plays an important role in animal welfare and animal husbandry (Dawkins 2003; Broom and Fraser 2015), as well as in conservation biology and biodiversity research. Often, findings from animal behaviour research have been picked up and oversimplified in popular science communications, thus it is important to have a critical eye when interpreting reported findings from popular science reports. Being able to refer to and understand the detailed findings of an original study are thus an important practice for animal researchers.

The methods summarised in this book are primarily intended to provide the tools to systematically investigate animal behaviour and thus make it objectively interpretable in a truly scientific fashion. In addition to fundamental considerations for scientific work, this book presents the methods of qualitative and quantitative recording of behaviour, considering both purely descriptive observations and experimental approaches. Even if some aspects of animal behaviour work may seem simple at first glance, such as observing animals, the systematic scientific practice is actually enormously complex. Behavioural biologists today often use highly specialized methods for data collection and analysis, such as automated sensors and tracking technologies, complex mathematical models, and artificial intelligence algorithms, as well as many other methods from other biological disciplines. However, well-planned data collection through personal obser-

1

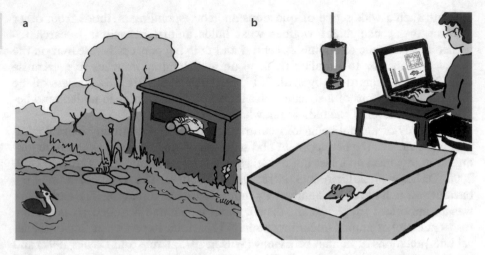

Fig. 1.2 Typical approaches for animal behaviour research include both (**a**) field work and (**b**) laboratory settings. (Drawings by Ulrich Pörschmann)

vation and experimentation, directly from animals in the field or in the laboratory (■ Fig. 1.2), continue to play a key role in animal behaviour research. Due to the multitude of factors to be considered, animal behaviour studies require special care in planning and execution. Knowledge of the more classical animal behaviour methods, i.e. how behaviour is registered and categorised and how experiments in animal behaviour are planned, carried out, evaluated and interpreted, is therefore still of great importance. Fundamental methodological knowledge is also crucial to be able to assess the quality and significance of animal behaviour studies e.g., when reading or reviewing papers.

1.2 Levels of Behavioural Analysis

In the different biological sub-disciplines, living organisms are studied at various systemic levels, ranging from research into interactions at the molecular level to the observation of complex ecosystems. Each of the biological sub-disciplines has an impact on the other disciplines to some extent. If we can understand behaviour in the context of an individual being faced with its environment, it becomes clear that there are various systemic levels at which animal behaviour research can be conducted (■ Fig. 1.3). At the organismic level, the questions usually address an entire animal, a group, or even a population. However, questions in animal behaviour can also refer to the internal processes within an animal. Questions about the relationships between behaviour and hormones, immune stress or neuronal processes, including learning, are part of animal behaviour research (e.g., Möstl and Palme 2002; Honarmand et al. 2010; McEwen 2020). Therefore, behavioural biology cannot (and should not) always be clearly distinguished from other biological fields.

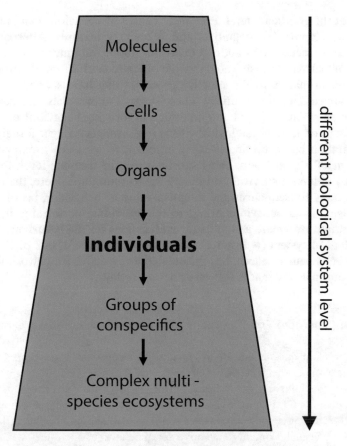

Fig. 1.3 Animal behaviour research can relate to the different biological system levels, with the individual level being the most central to the field

Depending on the research questions that are being addressed, the focus may be a single specific behaviour or a very complex behavioural strategy in which several behaviours are used in a targeted manner, such as complex courtship, aggression, or foraging behaviours. The behaviour of groups, such as swarm behaviour or behaviour during animal migration, can also be the focus of an investigation (e.g., Couzin et al. 2005). If more comparative aspects are involved, a higher system level, such as that of the population, animal species, or even a broader taxonomic clade, can be the subject of research (e.g., Bang and Cobb 1968; Reynolds and Szekely 1997).

The observation and research of animal behaviour phenomena can be generally grouped into two fundamental categories (▶ Box 1.1). On the one hand, questions can be asked that relate to the immediate factors that trigger behaviour, directly influence behaviour, underlie the control of behaviour, or lead to the achievement of an immediate goal. Such questions about *how* something works or has a direct effect, i.e. about the direct mechanisms of behaviour, are called questions at the **proximate level**. Proximate causes of behaviour can include physiological, developmental, cognitive, and hereditary features. Alternatively, animal behaviour research can address questions about the long-term, evolutionary functions of behaviour, also known as the **ultimate level**. Ultimate origins of behaviour pertain mainly to the selective processes that have shaped a behaviour's past and current functional utility. These distinct approaches can best be illustrated with an example. If you are interested in foraging behaviour of an animal, you may focus on how physiological effects (e.g., hunger) trigger foraging, or perhaps orientation based on an olfactory gradient, which leads proximately to the achievement of a goal (food, being satisfied). At the ultimate level, foraging behaviour can influence survival and the ability to reproduce. Here, the main focus is on the fitness consequences and adaptive value of behaviour, i.e. on the evolutionary advantage a behaviour brings to the individual performing it. **There are usually multiple proximate and ultimate explanations for the behaviours of animals in any biological system** (► Box 1.1). Niko Tinbergen (1963), one of the founders of animal behaviour research, has formulated a respective framework for observing behaviour in a way that is still very relevant today.

Box 1.1: Tinbergen's Four Questions and Proximate and Ultimate Explanations

Tinbergen's four questions relate to 1) the mechanisms that trigger/cause and control behaviour, 2) the ontogenesis of behaviour, i.e. how behaviour appears and changes throughout individual development, 3) the function, or adaptive value, of behaviour, and 4) its phylogenetic (evolutionary) origin. Every behaviour of an animal, whether it is, for example, bird song, animal migration, light–dark preferences of a unicellular organism, complex social behaviour, or strategies of foraging or avoiding predation, can be studied from all four perspectives, separately and in combination (◘ Fig. 1.4).

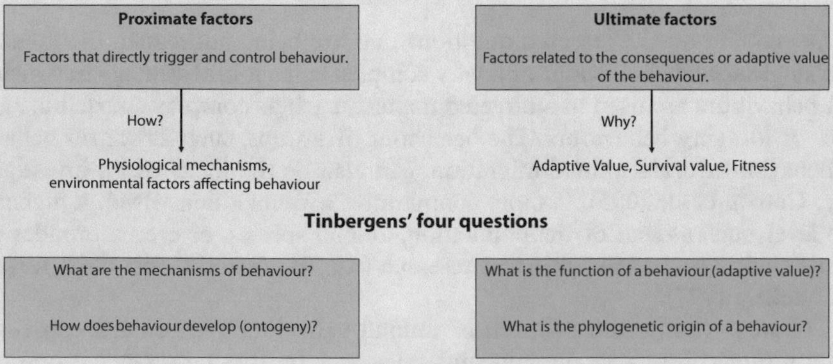

◘ **Fig. 1.4** Schematic representation of Tinbergen's four questions

It would be impossible to design a single study that addresses all four questions at the same time equally. Accordingly, it is important to separate the different levels at which research questions are addressed and to keep this in mind whenever planning a behavioural study.

Although the various aspects of behaviour cannot be addressed within a single study, it is generally true that the four questions suggested by Tinbergen (▶ Box 1.1) are equally important for understanding animal behaviour in all its dimensions. The mechanisms of behaviour play a crucial role in understanding the function of behaviour. To understand *why* an animal is doing something, it should also be known *how* perceived information is processed by the animal and what mechanistic possibilities are available to the animal. Moreover, when answering questions about the adaptive value of a behaviour, it must be considered that there may be behaviours that are not adaptive in a certain context. This can occur under a number of circumstances, for example when the underlying mechanisms have evolved in another context in which they are not at a disadvantage, and thus have not been selected against.

1.3 Anthropomorphism

In general, when measuring animal behaviour, care should be taken not to interpret the behaviour from a human perspective; the behaviour of animals should be described and interpreted with respect to their own biological background, without anthropomorphic interpretations. For example, if an animal shows little activity in an experiment, one should not describe it as "lazy" or even "tired", as these are internal states of an individual that are not accessible to us as human observers. Problems with anthropomorphism especially arise when it leads to interpretations of animal behaviour that cannot be explicitly verified by the data obtained. **The behaviour must be recorded in verifiable (see Chap. 2) and objective (see Chap. 4) measures.** Even if it is sometimes easier to describe the behaviour of an animal in the human sense, a distinction should be made between the scientific description (i.e. objectively verifiable) and a personal, subjective interpretation of the behaviour. However, a carefully employed anthropomorphisation can be helpful to communicate complex findings to a broader lay public, in order to draw attention and increase understanding of certain animal behaviours patterns (de Waal 2007), welfare issues or conservation concerns.

References

Alcock J (2013) Animal behavior: an evolutionary approach, 10th edn. Sinauer, Sunderland
Bang BG, Cobb S (1968) The size of the olfactory bulb in 108 species of birds. Auk 85:55–61
Broom DM, Fraser AF (2015) Domestic animal behaviour and welfare, 5th edn. CABI, Wallingford
Browne J (2013) Wallace and darwin. Curr Biol 23:R1071–R1072

Bierbach D, Landgraf T, Romanczuk P, Lukas J, Nguyen H, Wolf M, Krause J (2018) Using a robotic fish to investigate individual differences in social responsiveness in the guppy. Roy Soc Open Sci 5:181026

Byrne RW, Whiten A (Eds) (1988) Machiavellian intelligence: social expertise and the evolution of intellect in monkeys, apes, and humans. Clarendon Press/Oxford University Press, New York

Chouinard-Thuly L, Gierszewski S, Rosenthal GG, Reader SM, Rieucau G, Woo KL, Gerlai R, Tedore C, Ingley SJ, Stowers JR, Frommen JG, Dolins FL, Witte K (2017) Technical and conceptual considerations for using animated stimuli in studies of animal behavior. Curr Zool 63:5–19

Couzin ID, Krause J, Franks NR, Levin SA (2005) Effective leadership and decision-making in animal groups on the move. Nature 433:513–516

Danchin E, Giraldeau LA, Cezilly F (2008) Behavioural ecology. Oxford University Press, Oxford

Darwin C (1859) The Origin of Species by means of natural selection, or the preservation of favoured races in the struggle for life. John Murray, London

Davies NB, Krebs JR, West SA (2012) An introduction to behavioural ecology, 4th edn. Wiley Blackwell, Chichester

Dawkins R (1976) The selfish gene. Oxford University Press, Oxford

Dawkins MS (2003) Behaviour as a tool in the assessment of animal welfare. Zoology 106:383–387

Dugatkin LA (2019) Principles of animal behavior, 4th edn. University of Chicago Press, Chicago

Emery N (2016) Bird brain: an exploration of avian intelligence. Princeton University Press, New Jersey

Emery NJ, Clayton NS (2004) The mentality of crows: convergent evolution of intelligence in corvids and apes. Science 306:1903–1907

Honarmand M, Goymann W, Naguib M (2010) Stressful dieting: nutritional conditions but not compensatory growth elevate corticosterone levels in zebra finch nestlings and fledglings. PLoS ONE 5:e12930

Krause J, Winfield AF, Deneubourg JL (2011) Interactive robots in experimental biology. Trends Ecol Evol 26:369–375

Krebs JR, Davies NB (1997) Behavioural ecology – an evolutionary approach, 4th edn. Blackwell, Malden

McEwen BS (2020) Hormones and behavior and the integration of brain-body science. Horm Behav 119:104619

Möstl E, Palme R (2002) Hormones as indicators of stress. Domest Anim Endocrinol 23:67–74

Nicol CJ (2015) The behavioural biology of chickens. CABI, Wallingford

Reynolds JD, Székely T (1997) The evolution of parental care in shorebirds: life histories, ecology, and sexual selection. Behav Ecol 8:126–134

Tinbergen N (1963) On the aims and methods of ethology. Z Tierpsychol 20:410–433

Tomasello M, Call J (1997) Primate cognition. Oxford University Press, USA

De Waal F, Waal FB (2007) Chimpanzee politics: power and sex among apes. JHU Press.

Wilson EO (1975) Sociobiology: the new synthesis. Belknap, Cambridge

General Principles

Contents

2.1 Preliminary Considerations, Scientific Methodology, and Transparency

2

Observing the behaviour of animals so often leads to the questions: "What?", "How?", "Why?", "When?", "Who?" and "Where?". Through formulating such questions in a scientific manner, we generate hypotheses and predictions about their possible answers. Clear hypotheses, and the steps of scientific methodology based on them, are a basic foundation of scientific work (▶ Box 2.1).

Scientific studies in animal behaviour research are often time-consuming, or they might depend on particular seasons or locations and therefore cannot be completed or repeated in the short term. Thus, well-planned data collection is essential for valid and robust scientific work (▶ Box 2.1), and detailed and comprehensible documentation of the entire study, from start to finish, is required (Parker et al. 2016). With careful planning we can identify potential problems and pitfalls in advance, and we can prepare for well-structured and effective data acquisition, determine the sample size needed, as well as minimize undesirable interfering factors (also known as confounding factors). Here, we detail a number of aspects that are important to consider when planning a study.

Box 2.1: Scientific Methodology
Scientific work is based on several basic steps (◘ Fig. 2.1). These steps are summarized here in a schematic way, and later in the chapter we come back to them in more detail. First, research questions are based on observed events or suspected correlations, from this we formulate hypotheses and, in turn, make justified predictions. In experimental animal science, we plan experiments that are suitable for

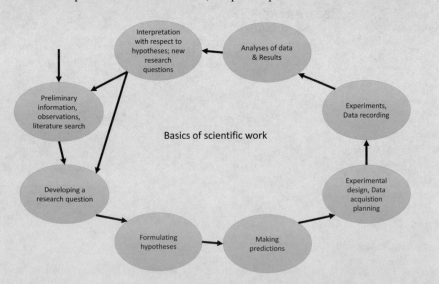

◘ **Fig. 2.1**   Rigid scientific work is essential for science and thus also animal behaviour research. The aim is to gradually narrow down the potential explanations by stepwise systematically excluding them

testing these hypotheses with their predictions. The results of the study are then statistically analysed according to our predictions. Finally, we decide whether our hypothesis can be maintained or whether the hypothesis is refuted (falsified).

We advise to discuss your data acquisition protocols/procedures and the experimental designs prior to the start of the study, for example with colleagues or in your institutional seminar. Considering different perspectives on a study design will help to optimize the design and to prevent potential pitfalls. You have to decide which data recording protocol is most suitable for the respective study and research question and which would cause the least problems in the subsequent interpretation of the results.

Careful planning of your experimental design and data acquisition in advance also avoids unnecessarily conducted experiments. Nothing is more annoying than to discover, after time-consuming and work-intensive data collection, that the data have little scientific value due to fundamental methodological faults. Moreover, in the case of experimental studies with animals, poorly planned studies can cause avoidable stress for the animals. Mistakes such as the lack of an appropriate control, imprecisely defined behaviours, unsuitable experimental design, or failure registration of important behaviours or environmental conditions during data collection can render a study obsolete and the results difficult or even impossible to interpret.

The statistical analysis of data is a central feature of scientific research, including animal behaviour research (e.g. Quinn and Keough 2002; Schielzeth and Nakagawa 2013; Gygax 2014; Jones et al. 2023). Even before starting a study, it is important to consider which statistical tests can be applied and what requirements they have with respect to data structure (e.g. paired or independent samples, random effects, e.g. Quinn and Keough 2002; Schielzeth and Nakagawa 2013), data distribution, or data type (e.g. categorial, numerical, ...). In line with this, one should also determine in advance what sample size is needed to answer the question (Faul et al. 2007). Such a priori sample size calculations (also called power analysis) are often required to register an experimental study according to the respective national laws.

Making your a priori planning of e.g. hypotheses, sample sizes, experimental design, and statistical methods transparent, either immediately or after an embargo, according to Open Science principles (Muñoz-Tamayo et al. 2022) can be worthwhile. Pre-registrations and/or registered reports are helpful tools, as they help to prevent questionable practices such as changing the statistical analysis after the experiments in order to get p-values below the significance threshold ("p-hacking") and adjusting hypotheses in accordance to the results ("HARKing", see below). In general, Open Science principles can be considered at every step of your scientific work to make your research accessible to everyone, improve reproducibility, and increase transparency (Muñoz-Tamayo et al. 2022). Open Science practices can include: i) pre-registration of your study, ii) publishing a registered report, iii) sharing your collected data, codes, etc., iv) publishing

2

preprints of your manuscripts, v) publishing your peer-reviewed papers in Open Access journals, vi) uploading the final versions of your papers to Open Access repositories and vii) participating in open peer review (Muñoz-Tamayo et al. 2022). You may not do all of these practices immediately or for every study, but each is becoming more common and it is advisable to consider these principles. Furthermore, Open Science principles can contribute to better animal welfare (Nawroth and Krause 2022). Web platforms such as e.g., Open Science Framework (▶ www.osf.io) support you making your Science open.

2.2 Research Questions, Hypotheses and Predictions

The first step in a research project is usually to formulate a research question. Do not underestimate the importance of the research question! It forms the foundation for the whole research process. Once you have established this solid foundation, it will naturally lead you to your hypotheses and predictions, which then determine how you collect your data, which determines how you can analyse your data, and finally, the results and their interpretation wrap up this coherent 'storyline' that you have created with your question at its root (◘ Fig. 2.1).

Starting with a clear research question, based on previous research, ensures that the findings of the investigation can be placed in a clearly interpretable biological context. When formulating a hypothesis for an experiment, it is essential that it is scientifically sound and can also be suitably tested with the experiment (▶ Box 2.2). To formulate a proper question and associated hypothesis, extensive knowledge of the study species and the theoretical background/literature is necessary (▶ Sect. 2.3). The formulation of hypotheses is also very important because it will force you to consider very carefully what expectations are placed on the planned data collection.

Box 2.2: Hypotheses
A hypothesis is a scientifically-based and verifiable statement. Hypotheses postulate connections between cause and effect in relation to observed phenomena. If possible, hypotheses contain only one testable variable and should be formulated in such a way that they are verifiable. Hypotheses can never be proven by a study, because alternative explanations can never be ruled out completely. Rather, hypotheses can only be rejected (falsified) or remain valid. By gradually excluding alternative hypotheses, through the scientific process, we can limit the number of possible explanatory hypotheses and come to the one that explains the behaviour of an animal best.

Hypothesis H1
The (working/alternative) hypothesis H1 is the hypothesis associated with the question and predicts that there is a difference between two or more samples.

Null hypothesis H0

The null hypothesis H0 states that there is no significant difference between two (or more) samples. The working hypothesis H1 is then tested against the null hypothesis with statistics (▶ Chap. 7). In practice, the null hypothesis is usually not explicitly specified because it is inherently opposed to the working hypothesis. An important prerequisite for formulating the null hypothesis is that it must be falsifiable, that is, that it can be refuted.

How must a hypothesis be formulated to be falsifiable?

Example of **the falsifiability** of a hypothesis using the example of the abundance of fire salamanders in a given habitat (◨ Fig. 2.2)

◨ **Fig. 2.2**   Picture of an adult fire salamander (*Salamandra salamandra*)

H1: There *are* fire salamanders in the specific forest area.
This working hypothesis H1 can be verified. However, it cannot be falsified, because in the case that no fire salamanders are found during an inspection of the forest, it may simply be that the forest has not been inspected carefully enough or at the wrong time.
H0: There *are no* fire salamanders in the specific forest area.
This null hypothesis H0, however, can be falsified. Once the first fire salamander is discovered, this hypothesis is falsified.

A clearly formulated hypothesis leads to exact predictions (▶ Box 2.1). Your predictions are specific statements of what you would observe if your hypothesis was

2

true, and are usually formulated with an if–then relationship. Predictions are a central point in hypothesis-oriented experimental research, as they provide a clear framework for data collection. This is where you decide which behaviours/physiological and ecological parameters etc. are recorded and what expectations you have for the outcome. In the case of predictions, it is already clear which interpretation possibilities are plausible if the hypotheses are fulfilled, or if they are not fulfilled.

By formulating your hypotheses and predictions in advance of data collection, you also actively prevent that you adjust the expectations of the experiment subsequently such that the results correspond exactly to the expectations (▶ Sect. 3.1.2). Indeed, you should never retrospectively adjust your hypothesis in accordance to the results of the study. This phenomenon is also known as 'HARKing' (Hypothesizing After Results are Known; e.g. Fraser et al. 2018) and is considered poor research practice. If an unexpected result emerges from your study, as it often does, the proper course of action would be to spell out this new idea and to conceive a new study that specifically tests the new hypothesis that arises from those findings.

2.3 Selection of the Study Species

The species on which you choose to carry out a particular project is usually based on one of two motives. On the one hand, your choice may be based on the biological question, and you find a suitable study species to address that question. On the other hand, your interest may be explicitly focused on a specific animal species. The motivation to learn more about a certain species can be a strong driving force behind scientific work. In many elaborate field studies that require enormous commitment, it is perhaps precisely this motivation that is the decisive force for conducting a scientifically demanding (long-term) study. An example of very long term studies on a convenient model organism is that of great tits in Wytham Woods, UK and in the Netherlands (e.g. Lack 1952; Krebs et al. 1978; Farine et al. 2015; van Oers and Naguib 2013; Bosse et al. 2017)

However, one must also keep in mind that a scientific study is more than collecting data and testing individual hypotheses. The findings of a (long-term) study should also be embedded in greater theory or be linked to an applied problem. This means, for example, that for a study which tests wider theories it should be established in advance to what extent new findings about the selected species may also be of general interest. If the results are only narrowly of interest for a specific species, the less generalisable the findings are, and therefore the less general interest there may be within the scientific community. There are of course exceptions to this, for example when considering nature conservation or other applied aspects such as husbandry conditions and animal welfare of farm or other captive animals. If you aim for the anticipated revelations from your study to be more generalised, it is important that you select an animal species that is suitable for a research question and research condition, and later explain this rationale in the introduction of your study. Often the decision which animal species is used for the study is based on a practical compromise. The more is already known about an animal species, the more detailed a study can be planned. Practical considerations can also play an important role in the selection of an animal species (► Box 2.3).

Box 2.3: Important Aspects in the Choice of Study Species (☐ Fig. 2.3)

☐ **Fig. 2.3** The selection of the animal species should fit the research question. Not every question can be answered with every animal species because the lifestyles and biology of the species often differ considerably, e.g. **a)** green-rumped parrot fish shown first and **b)** a polar bear (*Ursus maritimus*). (Photos by Oliver Krüger)

— Is the focus of your study on the specific species or on a specific research question?
— Is the study species suitable to answer your research question?
— How well can the knowledge gained from the study species be generalised?
— Is it practically possible to address this research question with the selected study species?

2.3.1 Domesticated Animals

Laboratories often keep domesticated animals, such as small mammals (e.g. rats, mice, guinea pigs), songbirds (e.g. domesticated zebra finches, canaries) or farm animals (e.g. chicken, turkeys, pigs, cattle) (► Box 2.4). These animals often offer practical advantages for scientific research and make it possible to investigate issues that are difficult or even impossible to address in the field. One advantage is that domesticated animals are easier to keep and handle than wild caught individuals (► Box 2.4), which are much more sensitive to laboratory conditions and experimental equipment than their domesticated counterparts. However, one should take into account to what extent the generalisability of the results obtained with a domesticated stock may be limited due to the domestication history. Although domesticated animals often show an almost completely natural

2

behavioural repertoire of the wild species (e.g. Jensen 1986), selective breeding for certain traits can lead to differences in behaviour between domesticated animals and the wild form (e.g. Tschirren et al. 2009; Krause and Schrader 2016). Such differences may lead to limitations in interpretation, especially when functional and evolutionary questions are to be addressed. Where animals have been bred for particularly extravagant traits such as colour, fin size or other morphological characteristics, or for specific behaviour or performance traits (egg count, weight gain) in non-natural social structures, the extent to which the results can be generalised should be carefully considered (▶ Box 2.4).

Box 2.4: What is Domestication?

Domestication is the process of describing changes in the phenotype (and thus also in behaviour) and genotype of animals that are caused by human selection and captive husbandry/rearing. In domesticated animals, natural selection has been replaced by artificial selection (Immelmann and Beer 1989). Three general factors have been described as central to domestication: 1) relaxation of natural selection (e.g. less predation pressure and ad libitum food); 2) intensive selection for traits that are desired by humans (e.g. appearance, performance); 3) there is also natural selection in captivity that leads to adaptations (Price 1997; Jensen 2006). Domestication leads, among other things, to changes in external morphology (such as size and coat/plumage colour; ◘ Fig. 2.4), internal morphology (such as reduction in brain size, differences in relative organ sizes), physiology (such as hormonal responses), development (such as earlier sexual maturity) and behavioural patterns (such as reduced anxiety) (Jensen 2006; Lesch et al. 2022).

◘ **Fig. 2.4** Example of how domestication has changed the plumage colour in Zebra Finches (*Taeniopygia guttata*). **a**) a wild coloured female and **b**) a domesticated leucistic female Zebra Finch selected for white plumage colour. Besides changes in plumage colour, domestication also has effects on other traits in domesticated populations of zebra finches (Forstmeier et al. 2007), suggesting that the relative relationship between traits in zebra finches has not been altered (Tschirren et al. 2009). (Modified from Hoffman et al. 2014, using the CC BY 4.0 license. ▶ https://doi.org/10.1371/journal.pone.0086519.g001)

2.4 Laboratory Tests and Field Work

Whether a study is carried out in the field or in the laboratory depends on the questions to be answered, the practical possibilities, as well as on personal interest and one's own abilities and possibilities. **Laboratory studies have the advantage that the experimental context can be carefully controlled.** In addition, stationary technical equipment can be used, which is often impractical or impossible in the field. Especially behavioural studies which require physiological measurements of animals, or which collect data from animals with large home ranges, can be quite limited in the field. The same applies to more complex experiments that require a stringent experimental design (▶ Box 2.5). Limitations of laboratory experiments, on the other hand, may be that the behaviour that an animal displays in a laboratory situation does not necessarily have to be of equal importance in its natural environment. Here, it should be considered to what extent results from laboratory experiments can be transferred to naturally occurring behaviours. In the field, it should be kept in mind that certain behaviours are likely to be affected by many factors acting simultaneously. It may also not be possible to study certain behavioural contexts in the field because they occur too rarely and are not accessible for data collection. The main **advantage of field research is that the behaviour of animals can actually be recorded in their natural environment and their natural context.** This is true for descriptive studies as well as field experiments. Reactions of animals to experimental stimuli in the field are easier to interpret with regard to the evolved function of the behaviour. **Due to the different advantages and disadvantages of both laboratory and field research, an ideal approach can be to combine the two.**

Box 2.5: Examples of Laboratory and Field Experiments
Example of a behavioural test under laboratory conditions
In a learning experiment with different domesticated chicken lines (*Gallus gallus domesticus*), the influence of the degree of selection on laying performance and on the abilities in an associative learning experiment was investigated. The learning was carried out in an automated skinner box with touch screen, on which the hens had to peck for their decisions (◘ Fig. 2.5). When the correct decision was made, the hens automatically received a food reward, controlled by a computer program. The experiment required a long, controlled training and testing phase, which is difficult to carry out under outdoor conditions. In addition, the test procedure was largely automated by technical equipment and software. In complete field conditions, such test equipment is more difficult to design (Dudde et al. 2018, 2022).

2

❏ **Fig. 2.5** Brown feathered laying hen (*Gallus gallus* domesticus) in automated Skinner box with touch screen. (From Dudde et al., 2018, using the CC BY 4.0 licence. ► https://doi.org/10.3389/fpsyg.2018.02000)

Example of a behavioural test in the field

In the field, the influence of early nutrition provided by parents on the stress response of the nestlings was investigated in great tits (*Parus major*) (❏ Fig. 2.6). The broods were cross-fostered (mutual exchange of offspring) to control for possible genetic effects. The parental supply of food was recorded with video cameras and analysed

❏ **Fig. 2.6**   Great tit (*Parus major*) in the field. (Photo by Oliver Krüger)

accordingly. At the age of two weeks, the nestlings were subjected to a "handling" test in the field, which is a good indicator of the personality of the young animals. All experiments were conducted in the field and demonstrated that parental food supply for young great tits is an important factor for the development of personality, which can be influenced by environmental conditions such as food availability (van Oers et al. 2015).

2.5 Scheduling a Research Project

As a rule, the time frame available for a study is almost always limited. Only a few weeks are available for short projects during the course of study, a few months for bachelor and master theses and only a few years for dissertations and many third-party funded projects. Even for long-term projects there can be logistical limitations, so **it is important to draw up a schedule in which the data is to be collected, analyzed and compiled.** Sufficient time buffers should be planned within the framework conditions, as delays can occur in the course of almost any study (▶ Box 2.6).

For field studies, bad weather is a particularly common problem that can lead to the suspension of data collection. For laboratory animals, acclimatisation in experimental set-ups may take longer than planned. Technical problems with video, audio or computer equipment occur more often than expected and desirable. It should also be taken into account that experiments or observations have to be repeated due to human error, for example equipment that has not been switched on or improperly set up. Time pressure resulting from a too tight schedule may lead to inaccurate data analysis or insufficient time to write up the study. In general, one should keep in mind that data analysis can often require more time than data collection.

Of course, schedules may change even after the start of a study, as circumstances may arise that were not considered in the initial planning. This makes it all the more important to keep track of how delays in trial preparation or data collection affect the overall schedule. A study that is not carefully planned in terms of time is always in danger of not being completed adequately.

Box 2.6: Time frame of a study
Various work steps must be taken into account in the time planning. It is important not to plan too tightly, but also to provide time buffers:
- Preparations (e.g. design of experiments, literature search, presentation of the plans in a seminar, practical work)
- Drafting of a data recording protocol (e.g. preparation of lists, preparation of software, determination of measurement parameters)
- Pilot phase (e.g. testing of the developed data acquisition protocols and test procedures), followed by revision of the protocols if necessary
- Data acquisition (e.g. test performance, observations).

2

- Data analysis (e.g. preparation of data in tables for subsequent analysis, statistical data evaluation with suitable tests and software such as R, creation of figures)
- Writing up the study (e.g. writing the study as a paper or final thesis, literature work, reference of the data to the question/ hypothesis, presentation of the results at conferences)

2.6 Marking of Animals for Individual Identification

For many animal behaviour questions it is a basic requirement that the observed animals can be individually distinguished. Natural characteristics of animals initially appear to be the most suitable method for individual recognition. Particularly mammals with individual fur patterns can often be reliably identified individually with appropriate practice (e.g. Bateson 1977; Hofer and East 1993). In marine mammals, features such as injuries on the fins are also used for individual recognition by observers (Wilson et al. 1997). Amphibians can also exhibit individual colour patterns that change over life but still allow differentiation from conspecifics (Drechsler et al. 2015; Goedbloed et al. 2017). If a study relies exclusively on natural characteristics, though, it can face limitations, especially in long-term studies or in studies on large groups of animals (e.g. Wolf et al. 2007; Meise et al. 2013). In addition, purely visual recognition can lead to that only a few, very experienced people are able to reliably recognise the animals. This can mean that the data cannot be verified by others and that new researchers in a group require a long training phase. For this reason, additional individual tagging is often useful, as far as this is technically possible, legal, and practical in terms of animal welfare or nature conservation considerations.

2.6.1 Marking Methods

Marking methods are diverse and vary with the animal species and with the issues involved (☐ Table 2.1). What is the aim of marking? How clear must the marking be? How long should the marking last? Do individuals have to be identifiable, or is it sufficient to assign them to specific groups? To what extent does the marking possibly restrict natural behaviour? **Each marking method has advantages and disadvantages.** In general, none of the possible marking methods can be described as the most suitable. **A method should always be selected to minimise its impact on the animal and data collection.**

An important factor in deciding how to mark an animal is the durability of the mark and its visibility during observation. If it is sufficient to identify the animal only during capture, more inconspicuous markings can be used than if the marking has to be visible during observation. Conspicuous markings can, for example, cause unintended effects on the social behaviour of animals (e.g., Burley 1986; Frommen et al. 2015) or may alter predation risk for the marked individuals (e.g., Zuberogoitia et al. 2012, Calderon-Chalco and Putnam 2019).

◻ **Table 2.1** Examples of marking methods for different groups of animals. In some cases, different methods can also be combined

Marking method	Animal species	Advantages	Disadvantages
Numbered (aluminium) rings	Birds	Permanently	Not readable from a distance
Colour rings/colour collars, also with numbers (◻ Fig. 2.8)	Birds/Mammals	Durable, easily visible from a distance	Limited number of combinations, possibly not weatherproof, may affect behaviour
Wing marks/fin marks (◻ Fig. 2.9)	Birds, mammals	Clearly visible depending on the size of the inscription	Depending on size and position possibly disturbing
Branding	Mammal	Permanent	Painful for the animal, inflammation possible, irreversible
Colour markings (◻ Fig. 2.7)	Many taxa	Easily recognisable	Possibly not permanent, can disturb behaviour
Ear tags (◻ Fig. 2.7), wing tags	Mammals/Birds	Permanent	Depending on size, not readable from a distance, possibly disturbing
Fur cuts	Mammal	Low stress	Temporary
Passive (RFID) transponders (◻ Fig. 2.7)	Many taxa	Permanent	Short reach
Active transponders/ radio or gps tags	Many taxa	Long range	Limited service life due to batteries unless solar powered
Elastomers (◻ Fig. 2.7)	Fish	Easily recognisable	Limited number of colour combinations

The use of small **RFID transponders** (RFID = Radio-Frequency Identification) is a comparatively elegant method of animal marking, since the marking usually does not disturb the animal and is hardly or not visible to other animals (◻ Fig. 2.7). RFID transponders are small passive transmitters whose individual number code can be read by readers/antennas from close range, e.g. when animals enter or pass through a particular place. Transponders have the further advantage that, like the classic aluminium rings used in bird ringing, they have a unique number code. This allows large numbers of animals to be marked, while at the same time ensuring that the individual marking lasts for long periods of time. The disadvantages of using transponders are that they are not suitable for all animal species, as they may not be able to be attached or may not be practical due to the short range of the signal (< 0.5m). The implantation of transponders (e.g. in small mammals) may require the animals to be anesthetised, so, as with all marking methods, it must be considered whether the associated intervention can be justified scientifically and in terms of animal welfare.

2

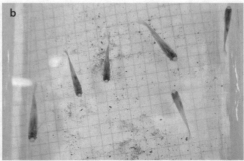

◘ Fig. 2.7 Examples of markings on individual animals: a) Domestic pig (*Sus scrofa* domesticus) with two ear tags and a colour marking on the back. The left ear tag as seen from the animal is an ear tag which may have a printed number. In addition to an imprint, the tag in the right ear contains an RFID transponder with which, for example, the whereabouts of the animal can be recorded automatically using suitable antennas. The ear tags represent permanent markings, whereas the coloured marking on the back is only visible for a few days (Photo by Antonia Patt), b) Guppies (*Poecilia reticulata*) with Visible Implant Elastomers (VIE tags). (Photo by Christopher Schutz)

The **aluminium or coloured rings** used in bird ringing (◘ Fig. 2.8) have similar advantages and disadvantages to transponders, for example that the animals must first be caught. With practice and patience, coloured rings can then be read from a distance using binoculars or a spotting scope. Furthermore, well attached leg rings rarely have a negative effect on the birds (but see Griesser et al. 2012). When using colour rings, however, it should be considered if specific colours may affect the behaviour of the individuals or of others towards them (Burley 1986; Pariser et al. 2010; Seguin and Forstmeier 2012).

Animals observed in long-term studies are often permanently marked individually (◘ Fig. 2.9). Here, as with short-term marking, care should be taken to ensure that the markings do not obstruct the animal and that they cannot be removed or manipulated by the animal itself or by other animals. This means that the markings should be as behaviour-neutral as possible. It is also important to note that the markings should be suitable in the environmental conditions, i.e., as in the case of Galapagos sea lions (Wolf et al. 2007; Meise et al. 2013; ◘ Fig. 2.9), they should be robust against strong sunlight, high temperatures, and salt water.

Another individual marking method is the use of **radio transmitters, geolocators, or satellite transmitters**, which can be used to locate and determine animals individually from a distance and with which the spatial movements of the animals can also be recorded (e.g. Cagnacci et al. 2010; Brown et al. 2012; Snijders et al. 2017). Such transmitters are attached externally to the animal. Traditionally they have a limited life span due to their battery operation, although solar-powered and hybrid (solar-charging) options have been recently developed. Radio transmitters are suitable for animals in dense habitats where direct observation is limited, or for nocturnal animals (if not fully solar powered). Geolocators and satellite telemetry are used especially for large-scale animal migration. Since radio telemetry transmitters in particular usually have a limited period of

☐ **Fig. 2.8** Examples of markings on individual animals: a) Chicken (*Gallus gallus* domesticus) with colour rings applied. The colour rings are relatively easy to see with clear and strong colours, but the number of possible colour combinations is quickly a limiting factor. For longer observation periods, it should be noted that colours or markings may change or disappear due to weathering and light. b) A male zebra finch (*Taeniopygia guttata*) with a plastic ring on its left leg. (Modified from Hoffman et al. 2014, using the CC BY 4.0 license. ► https://doi.org/10.1371/journal.pone.0086519.g001)

2

◘ **Fig. 2.9** Examples of markings on individual animals: Galapagos sea lion (*Zalophus wollebaeki*) with small coloured and numbered mark on the fin. Marked animals can be easily recognized and the numbers are easy to read with binoculars or close up. As with all labelled markings, one has to find an optimum size of the marking so that on the one hand the marking is easy to read and on the other hand the marking does not become too large. (Photo by Oliver Krüger)

Appropriate animal husbandry for research purposes

Optimal compromise from different perspectives, taking into account the applicable legal basis

Demands on animal husbandry from the perspective of **laboratory animals**:	Demands on animal husbandry from the perspective of **animal care workers**:	Demands on animal husbandry from the perspective of **researchers**:
- A natural environment - Adequate nutrition - Performance of the natural behavioural repertoire - Social interactions - Avoidance of pain, suffering, fear, stress - Freedom of movement - Reproduction - …	- functional husbandry - Work safety - Working hours - Functional distribution of the animals in the rooms/cages (good management) - Hygiene requirements - Economical work (consumables, personnel) - Animal health - Animal welfare - …	- Standardization of the trials - Spatial proximity to the experimental rooms - Unrestricted access to the animals - Unbureaucratic work (purchasing, breeding, applications) - Detailed documentation - Animal health - Animal welfare - Working hours - Occupational safety - …

◘ **Fig. 2.10** Requirements for animal husbandry as a compromise of three different fields of interest, which have to be united within the legal framework

activity, animals equipped with transmitters are often additionally provided with more permanent markings.

In principle, markings that lead to measurable changes in the behaviour of the animals or have an influence on their survival should be avoided. In addition, any stress to the animal when applying the markings must be minimised. Finally, it should be carefully considered which individuals are marked; for example, in the case of chicken in existing groups, it can be problematic to mark only part of the group (Dennis et al. 2008).

2.6.2 Naming of Animals

Especially in studies on certain species, there is a tendency to name animals with names taken from human language. This approach is particularly common for long-lived animals in zoos or research institutes (e.g. Pepperberg 2009; Manrique et al. 2013). Names can have the advantage that they are easier to remember than colour or number codes. However, names should not be chosen carelessly. Names that may have an association with certain experiences of the researchers have the potential to create a bias or expectation. Therefore, naming animals using neutral codes is often more scientific, since no subjective assessments are associated with the codes themselves. In databases, it is also easier to encode age or kinship relationships in numbers or letter codes, which can be advantageous for data analysis. It is therefore recommended to use names only where the animals have already been named by others, where such a designation system is already established, or under a specific, objective protocol.

2.7 Animal Husbandry

When animal behaviour studies are carried out under laboratory conditions, it is necessary to keep animals in cages, aquariums, terrariums, stables, runs or aviaries. There are many aspects to consider in animal husbandry to ensure the welfare and health of the animals and the success of the experiments (◘ Fig. 2.10). There are national legal requirements that must be adhered to whenever keeping animals. The specifics differ by country, but most generally require that an animal can be fed and cared for appropriately according to its species-specific needs, and that it is housed in a manner appropriate to its behaviour. Animal husbandry must not restrict the animals' ability to move to such an extent that pain or avoidable suffering or damage is inflicted on the animal.

There are various fields of interest in the planning and maintenance of animal husbandry (Fig. 2.10), which must all be optimally combined within the framework of the applicable national laws and regulations. First of all, a distinction must be made between standard animal husbandry, in which animals are kept long term, and short-term changes in husbandry for experimental purposes. In addition to the experimental demands on husbandry, practicability and good care facilities also play an important role in long-term animal husbandry. Furthermore, the lab-

2

oratory animals themselves have an important interest in the conditions of husbandry that they cannot communicate directly to us. In order to ensure appropriate animal husbandry with good animal welfare and good animal health, the demands of all three fields of interest of laboratory animals, animal care, and those responsible for the experiments should be combined to form the best possible compromise (◙ Fig. 2.10). There is no patent solution for optimal animal husbandry, since the requirements depend on the animal species, the respective spatial and personnel conditions and constraints resulting from the experimental design. It is therefore important to estimate in each individual case how animal husbandry can be optimised within the framework of these requirements. Accordingly, the use of husbandry systems that are very restrictive for the animals or so complex and expensive that they require many resources should be justified. Such aspects should ultimately be considered when choosing the species (▶ Sect. 2.3) for a scientific study.

The short-term restrictive housing of animals for experimental purposes may be assessed differently than the long-term housing conditions. In behavioural studies in which it is important that animals are allowed to move as free of stress as possible, e.g. in choice or learning experiments with positive rewards, it should be taken into account in husbandry that the capture and transfer of animals to other social groups or to an individual enclosure necessary for data collection can be a stress factor (Kaiser and Sachser 2005; Sachser et al. 2011). Accordingly, it should be estimated in advance how long an animal needs to get used to new housing conditions in order to show sufficient activity or exploration for the experiment and to resume its normal behaviour.

2.8 Ethical Aspects of Working with Animals

Researchers conducting studies with animals take on a responsibility to treat the animals with respect and care (Dawkins 1980, 2006; Bateson 2005; Hubrecht 2014). No one should inflict pain, suffering or harm on an animal without reasonable cause and strong justification.

At the legal level, a distinction is made between animal experiments that require authorisation, animal experiments that must be reported, and studies on animals that are not covered by legal protections. Which authorisation is required must be clarified before the experiment is started in accordance with the respective national law. It is important to clarify in advance whether and which permits are necessary for a study. Since official approval procedures can take time, an application for approval should be submitted accordingly long before the planned start of the experiment. In principle, approval should be granted by the relevant competent authority before any experiment begins. According to most guidelines, animal experiments already begin with comparatively minor invasive procedures, such as the taking of blood for scientific rather than medical diagnostic purposes. Experiments on invertebrates (with the exception of cephalopods) and non-invasive experiments on vertebrates are often not subject to authorisation. However, depending on the context of the studies, purely behavioural experiments may also be subject to authorisation. In case of doubt, the competent authorities should be consulted in advance of a planned study.

Particularly in the case of invasive animal experiments, questions of ethical acceptability arise. Considerations of the ethical justifiability of one's own animal experiments should not only be seen as a mandatory task for the submission of an application for an animal experiment, but also serve to weigh up whether the experiment is correctly planned and necessary: How many laboratory animals are really necessary? Can better and gentler methods be used? What is the gain in knowledge from the study and is it in relation to the animal experiment? Can the distress of the animals in the experiment be reduced? These considerations are of central importance in order to fulfil an ethical obligation to the animals and also to be able to justify the investigations to society.

In such considerations and in order to meet ethical requirements, the **3Rs principle** of Russell and Burch (1959) should be taken into account, especially in medical-pharmacological experiments. This principle aims to promote the **reduction** of the number of individuals, **refinement** of the methodology, and the development and use of **replacement** and supplementary methods of animal experiments and their methods (◘ Fig. 2.11). Despite the fundamental central importance of the three Rs, they do not always apply in the same way in animal behaviour studies. For example, using an alternative organism as a research model would not be an option if the focus of the investigation is the behaviour of a specific animal species. Nevertheless, it is absolutely necessary and sensible, especially in experiments that cause pain or suffering to the animals, to reduce the number of animals involved to a minimum (Ruxton 1998) while still having sufficient replicates to answer the scientific question (which may be quantified with power analyses), and to weigh the costs for the animal responsibly against the expected gain in scientific knowledge.

In summary, it is critically important that animal experiments are ethically and legally justifiable. A careful balance must be found between any suffering of the study animals and the derived results, the gain in knowledge, and their significance for scientific progress.

2.8.1 Animal Welfare

An important aspect of working with animals is to monitor animal welfare. **A primary problem in measuring animal welfare is to find suitable parameters that can be indicators of animal welfare.** It is not possible to ask animals directly about their welfare. Instead, we can "ask" animals to make choices and obtain a weighting of preferences, which is only meaningful in relation to the choices made. An initial approach to assessing animal welfare has long been based on **resource-centric indicators**, such as how much space the animals have available, how the enclosure is designed, and whether the animals have enough appropriate food. With this approach, good husbandry can be achieved. However, the actual individual animal welfare can only be assessed using **animal-centric indicators.** One approach to assessing animal welfare taking into account the needs of the animals is

2

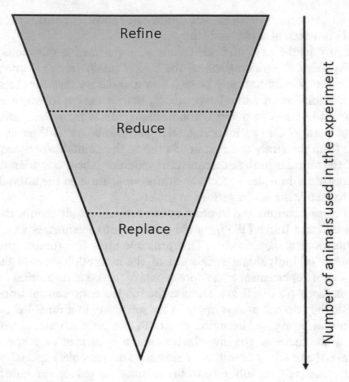

◘ Fig. 2.11 Possibilities of reducing the number of animals tested in experiments using the 3Rs principle according to Russell and Burch (1959). The aim is to reduce the number of animals, to refine methods—which also means fewer animals are needed—and to replace animal experiments with alternative and supplementary methods. Society and scientists have substantially increased their efforts over the years to provide better information about animal experiments

the **principle of five freedoms** developed in farm animal science (FAWC 2010; McCulloch 2013): 1) Freedom from hunger and thirst, 2) Freedom from discomfort through a good housing environment, 3) Freedom from pain, injury and disease, 4) Freedom to behave normally, 5) Freedom from fear, stress and suffering. Based on these five freedoms, an attempt was made to combine resource-centric animal welfare indicators with animal-centric welfare indicators. In addition to animals in laboratories, the assessment of animal welfare also plays a special role for farm animals and in wildlife. Consequently, there are many different approaches to measuring animal welfare.

References

Bateson PPG (1977) Testing an observer's ability to identify individual animals. Anim Behav 25:247–248

Bateson P (2005) Ethics and behavioural biology. Advances in the Study of Behavior 35:211–233

Bosse M, Spurgin LG, Laine VN, Cole EF, Firth JA, Gienapp P, Gosler AG, McMahon K, Poissant J, Verhagen I, Groenen MAM, van Oers K, Sheldon BC, Visser ME, Slate J (2017) Recent natural selection causes adaptive evolution of an avian polygenic trait. Science 358:365–368

Brown DD, LaPoint S, Kays R, Heidrich W, Kümmeth F, Wikelski M (2012) Accelerometer-informed GPS telemetry: reducing the trade-off between resolution and longevity. Wildl Soc Bull 36:139–146

Burley N (1986) Sex-ratio manipulation in color-banded populations of zebra finches. Evolution 40:1191–1206

Cagnacci F, Boitani L, Powell RA, Boyce MS (2010) Animal ecology meets GPS-based radiotelemetry: a perfect storm of opportunities and challenges. Philos Trans R Soc B 365:2157–2162

Calderon-Chalco KA, Putman BJ (2019) The effect of paint marking on predation risk in western fence lizards: a test using clay models. Herpetol Conserv Biol 14:80–90

Dawkins MS (1980) Animal suffering: the science of animal welfare. Chapman and Hall, London

Dawkins MS (2006) A user's guide to animal welfare science. Trends Ecol Evol 21:77–82

Dennis RL, Newberry RC, Cheng HW, Estevez I (2008) Appearance matters: artificial marking alters aggression and stress. Poult Sci 87:1939–1946

Drechsler A, Helling T, Steinfartz S (2015) Genetic fingerprinting proves cross-correlated automatic photo-identification of individuals as highly efficient in large capture–mark–recapture studies. Ecol Evol 5:141–151

Dudde A, Krause ET, Matthews LR, Schrader L (2018) More than eggs – relationship between productivity and learning in laying hens. Front Psychol 9:2000

Dudde A, Phi-Van L, Schrader L, Obert AJ, Krause ET (2022) Brain gain—Is the cognitive performance of domestic hens affected by a functional polymorphism in the serotonin transporter gene? Front Psychol 13:901022

Farine DR, Firth JA, Aplin LM, Crates RA, Culina A, Garroway CJ, Hinde CA, Kidd LR, Milligan ND, Psorakis I, Radersma R, Verhelst BL, Voelkl B, Sheldon BC (2015) The role of social and ecological processes in structuring animal populations: a case study from automated tracking of wild birds. Royal Society Open Science 2:150057

Faul F, Erdfelder E, Lang AG, Buchner A (2007) G* Power 3: A flexible statistical power analysis program for the social, behavioral, and biomedical sciences. Behav Res Methods 39:175–191

FAWC (2010) Annual review 2009–2010. Farm Animal Welfare Council. UK

Forstmeier W, Segelbacher G, Mueller J, Kempenaers B (2007) Genetic variation and differentiation in captive and wild zebra finches (*Taeniopygia guttata*). Mol Ecol 16:4039–4050

Fraser H, Parker T, Nakagawa S, Barnett A, Fidler F (2018) Questionable research practices in ecology and evolution. PLoS ONE 13:e0200303

Frommen JG, Hanak S, Schmidl CA, Thünken T (2015) Visible implant elastomer tagging influences social preferences of zebrafish (*Danio rerio*). Behaviour 152:1765–1777

Goedbloed JD, Segev O, Küpfer E, Pietzsch N, Matthe M, Steinfartz S (2017) Evaluation of a new Amphident module and sources of automated photo identification errors using data from *Salamandra infraimmaculata*. Salamandra 53:314–318

Griesser M, Schneider NA, Collis MA, Overs A, Guppy M, Guppy S, Takeuchi N, Collins P, Peters A, Hall ML (2012) Causes of ring-related leg injuries in birds–evidence and recommendations from four field studies. PLoS ONE 7:e51891

Gygax L (2014) The A to Z of statistics for testing cognitive judgement bias. Anim Behav 95:59–69

Hofer H, East ML (1993) The commuting system of Serengeti spotted hyaenas: how a predator copes with migratory prey, I. Soc Organ Anim Behav 46:547–557

Hoffman JI, Krause ET, Lehmann K, Krüger O (2014) MC1R genotype and plumage colouration in the zebra finch (*Taeniopygia guttata*): population structure generates artefactual associations. PLoS ONE 9:e86519

Hubrecht RC (2014) The welfare of animals used in research: practice and ethics. Wiley, Ames

Immelmann K, Beer C (1989) A dictionary of ethology. Harvard University Press, Cambridge

Jensen P (1986) Observations on the maternal behaviour of free-ranging domestic pigs. Appl Anim Behav Sci 16:131–142

Jensen P (2006) Domestication – From behaviour to genes and back again. Appl Anim Behav Sci 97:3–15

Jones E, Harden S, Crawley MJ (2023) The R book, 3rd edn. Wiley, Hoboken

Kaiser S, Sachser N (2005) The effects of prenatal social stress on behaviour: mechanisms and function. Neurosci Biobehav Rev 29:283–294

Krause ET, Schrader L (2016) Intra-specific variation in nest-site preferences of zebra finches: do height and cover matter? Emu-Austral Ornithology 116:333–339

Krebs J, Ashcroft R, Webber M (1978) Song repertoires and territory defence in the great tit. Nature 271:539–542

Lack D (1952) Reproductive rate and population density in the great tit: Kluijver's study. Ibis 94:167–173

Lesch R, Kotrschal K, Kitchener AC, Fitch WT, Kotrschal A (2022) The expensive-tissue hypothesis may help explain brain-size reduction during domestication. Commun Integr Biol 15:190–192

Manrique HM, Völter CJ, Call J (2013) Repeated innovation in great apes. Anim Behav 85:195–202

McCulloch SP (2013) A critique of FAWC's five freedoms as a framework for the analysis of animal welfare. J Agric Environ Ethics 26:959–975

Meise K, Krüger O, Piedrahita P, Trillmich F (2013) Site fidelity of male Galápagos sea lions: a lifetime perspective. Behav Ecol Sociobiol 67:1001–1011

Nawroth C, Krause ET (2022) The academic, societal and animal welfare benefits of open science for animal science. Front in Vet Sci 9:810989

Muñoz-Tamayo R, Nielsen BL, Gagaoua M, Gondret F, Krause ET, Morgavi DP, Olsson IAS, Pastell M, Taghipoor M, Tedeschi L, Veissier I, Nawroth, C (2022) Seven steps to enhance open science practices in animal science. PNAS Nexus 1:pgac106.

Pariser EC, Mariette MM, Griffith SC (2010) Artificial ornaments manipulate intrinsic male quality in wild-caught zebra finches (*Taeniopygia guttata*). Behav Ecol 21:264–269

Parker TH, Forstmeier W, Koricheva J, Fidler F, Hadfield JD, Chee YE, Kelly CD, Gurecitch J, Nakagawa S (2016) Transparency in ecology and evolution: real problems, real solutions. Trends Ecol Evol 31:711–719

Pepperberg IM (2009) The Alex studies: cognitive and communicative abilities of grey parrots. Harvard University Press, Cambridge

Price EO (1997) Behavioural genetics and the process of animal domestication. In: Grandin T (ed) Genetics and the behaviour of domestic animals. Academic, pp 31–65

Quinn GP, Keough MJ (2002) Experimental design and data analysis for biologists. Cambridge University Press, Cambridge

Russell WMS, Burch RL (1959) The principles of humane experimental technique. Methuen, London

Ruxton GD (1998) Experimental design: minimizing suffering may not always mean minimizing the number of subjects. Anim Behav 56:511–512

Sachser N, Hennessy MB, Kaiser S (2011) Adaptive modulation of behavioural profiles by social stress during early phases of life and adolescence. Neurosci Biobehav Rev 35:1518–1533

Schielzeth H, Nakagawa S (2013) Nested by design: model fitting and interpretation in a mixed model era. Methods Ecol Evol 4:14–24

Seguin A, Forstmeier W (2012) No band color effects on male courtship rate or body mass in the zebra finch: four experiments and a meta-analysis. PLoS ONE 7:e37785

Snijders L, Weme LEN, de Goede P, Savage JL, van Oers K, Naguib M (2017) Context-dependent effects of radio transmitter attachment on a small passerine. J Avian Biol 48:650–659

Tschirren B, Rutstein AN, Postma E, Mariette M, Griffith SC (2009) Short- and long-term consequences of early developmental conditions: a case study on wild and domesticated zebra finches. J Evol Biol 22:387–395

van Oers K, Kohn GM, Hinde CA, Naguib M (2015) Parental food provisioning is related to nestling stress response in wild great tit nestlings: implications for the development of personality. Front Zool 12:S10

van Oers K, Naguib M (2013) Avian personality. In: Carere C, Maestripieri D (eds) Animal personalities: behavior, physiology, and evolution. University of Chicago Press, Chicago, pp 66–95

Wilson B, Thompson PM, Hammond PS (1997) Habitat use by bottlenose dolphins: seasonal distribution and stratified movement patterns in the Moray Firth, Scotland. J Appl Ecol 34:1365–1374

Wolf JBW, Mawdsley D, Trillmich F, James R (2007) Social structure in a colonial mammal: unravelling hidden structural layers and their foundations by network analysis. Anim Behav 74:1293–1302

Zuberogoitia I, Arroyo B, O'Donoghue B, Zabala J, Martínez JA, Martínez JE, Murphy SG (2012) Standing out from the crowd: are patagial wing tags a potential predator attraction for harriers (Circus spp.)? J Ornithol 153:985–989

Experimental Planning and Experimental Design

Contents

3.1 Descriptive and Experimental Research

The various considerations that must be taken into account before embarking on an animal behaviour study differ depending on whether the study is descriptive or experimental. In descriptive or correlative studies, the naturally occurring behaviour of animals is observed without influence from the researcher. In an experimental study, an experimenter tests animals under controlled conditions, typically in an environment in which the animals are exposed to specific stimuli, tasks, or conditions. Descriptive and experimental studies both use quantitative methods to measure and assess animal behaviour, and both approaches can be carried out in the laboratory or in the field. However, the conclusions that can be drawn from descriptive and experimental studies fundamentally differ (◘ Table 3.1). In a descriptive study, the relationship between a behaviour and a context can be analyzed through correlations, but we cannot determine a cause-and-effect relationship. The reason is that any correlation between a behaviour and a context can be driven by a third variable that affects both at the same time. If an animal is more active when the sun is shining, it can be that the sun indeed directly causes the animal to move, yet it is also possible that the sun triggers the movement of prey species, which in turn triggers the movement of the focal animal. To determine *causation,* experiments are necessary in which the factor to be investigated can be specified via appropriate experimental and control conditions (◘ Table 3.1).

To design experiments that address specific scientific questions, it is critical to have some knowledge of the natural correlations, as they can be demonstrated in descriptive studies. Experiments in which animals are studied in contexts that have little to do with the natural context could reveal relationships that do not exist in nature. In this sense, research projects in which both descriptive and experimental work is integrated are particularly insightful (e.g. Krüger and Lindström 2001; Chakarov et al. 2017; Loning et al. 2022).

3.1.1 Preliminary Observations and Pilot Studies

Before we start any quantitative data collection, it is best to do preliminary observations, literature searches and/or pilot studies. Such pre-observations usually

◘ **Table 3.1**   Key characteristics of descriptive and experimental research. There are good reasons for both approaches, and the combination of both is a particular strength of scientific research. From descriptively collected data sets it is not easy to draw conclusions about cause-effect relationships between observed variables. In an experimental laboratory study, limitations are more likely to be found in estimating the extent to which the data collected are also relevant to the natural context

Descriptive	Experimental
Natural or captive context	Experimental control
Determine natural correlations	Determine causal relationships
Causal relationships cannot be established or can only be determined to a limited extent	
Field and/or captivity	Laboratory and/or field

provide the basis to develop more relevant, specific questions and to generate appropriate hypotheses. **Ample knowledge of the natural behaviour of the species facilitates formulating questions more specifically, planning how to record the data and it minimises potential problems in the process.** Studying relevant original literature is an essential step in this process. Further, with a pilot study, above all, we can test the data recording methods (▶ Chap. 4) and adapt the experimental conditions. We then can also check whether the planned categorisation of behaviour during an observation is as feasible as intended. It is important to keep in mind that also pilot studies may require specific ethical permits.

A pilot study can be important to practice handling the animals, the equipment, the experimental set-up, or observation methods. Observing animals in the field and documenting the observations effectively, so that the data can be quantitatively analysed, requires practice and can be more demanding in reality than expected. During the pilot it is thus important to determine which behaviours need to be recorded to answer the research questions. It thus is also important to test and document the data recording methods carefully and such that they can be repeated by other observers (▶ Chap. 5). Descriptions based on counts or measurements tend to be the most objective (▶ Box 3.1). A pilot study should determine which and how many behaviours can be recorded simultaneously. In addition, it can reveal whether individuals can be easily identified individually. A pilot study will reveal whether the intervals of data registration (▶ Chap. 4) are too short to accurately record the observations, or whether the intervals are too long and thus risk missing important information about the relevant behaviour.

Box 3.1: Distinction Between Interpretive and Purely Descriptive Registration of Behaviour

◻ **Fig. 3.1**   Leaf-cutter ants. (Photo by Oliver Krüger)

3

It is best to record the data in such a way that they are not subjectively interpreted, but objectively verifiable. Objective data records are usually based on counts or measurements.

We may observe, for example, that an insect has six legs (■ Fig. 3.1) and is 5 cm long, or that an animal stays in a certain area for a measured duration. More subjective descriptions of observations are those that are observer-specific and thus cannot be verified by others. For example, a person describes an insect as "small" and "fast moving" or without accurate time measurements that an animal has stayed longer in one area of an experimental set-up than in another area. However, "small" and "fast" do not have the same meaning for every observer, and durations can also be perceived differently by different observers, so that such anecdotal descriptions have little scientific value. In contrast, the description that an insect runs 10 cm/s or that it has spent 30 min in one experimental area and 10 min in another is more objective and can be verified.

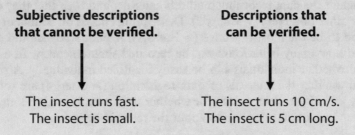

An important part of a pilot study is also to assess in advance whether the data structure is such that one can analyse it with suitable statistical methods. Especially in the case of complex data structures with e.g. nested designs or repeated measurements, it is important to think in advance about the statistical analysis. A pilot study in the field can reveal whether written or voice notes allow a quantitative analysis. Often one becomes aware of the value of field notes only when transferring them numerically to columns in tables. Likewise, a pilot analysis of audio or video recordings will help to decide if the data collection indeed should be carried out as planned. Often it only becomes clear how the required data need to be recorded once the analysis is carefully considered. Postponing consideration of the analysis to the moment when the final analysis needs to be done can lead to avoidable analytical problems. **Taken together, pilot studies can significantly contribute to the success of a study. Only serious testing will allow us to become confident that the planned steps of data collection can be implemented and that the recorded data will be analysable as planned.**

3.1.2 Descriptive Data Recording

Descriptive studies are in some ways more susceptible to not being planned in as much detail as experimental studies. Experimental studies tend to require more concrete practical considerations, such as designing an experimental set-up. In descriptive studies, in contrast, one can be more inclined to start data collection unprepared. Such an approach has at least two dangers of working unscientifically. On the one hand, the protocol according to which one is recording the data may not have the structure required for later statistical analysis. On the other hand, without clear preliminary considerations, one risks searching in the data after they have been collected, until data patterns are found that appear interesting. Such explorative data analyses, i.e. data analyses without prior clear hypotheses, are fine, but must be clearly identified as such. They must not lead to hypotheses being claimed after the analyses were performed so that these hypotheses then are being tested with the same data which were used to generate them. This is an unacceptable circular practice, also known as HARKing (Hypothesizing After Results are Known; Kerr 1998).

If a descriptive study is well planned, it is determined in advance what the aim is of the data collection. For which questions do we want to collect the data and what conclusions do we want drawn from the data? A common misconception is that descriptive studies can be used to draw conclusions about causal relationships (▶ Sect. 3.1.1).

Like for any study, in a descriptive study it is also important to create a systematic data recording protocol at the beginning. Here, we can then enter our observations and note down the meta-data such as time, place, weather, context, etc. As mentioned above, the behaviours we want to record need to be clearly defined so that we can record their occurrence, and if applicable their start and end time. Prior to starting we of course also need to have established how to record the behaviour, i.e. the timeline and the level of recording detail (▶ Chap. 4). In addition, depending on the research question, descriptive data collection may also require control conditions, in the same way as is necessary for experiments. In principle, most of the aspects that need to be considered for an experimental study also apply to descriptive research.

3.2 Planning of Experiments: Experimental Groups and Data Structure

In the simplest case of experimental research on animal behaviour, one variable is manipulated in order to determine its influence on a behaviour. This experimentally controlled variable is called the **independent variable** and is defined by the person conducting the experiment. Independent variables are the experimental conditions (treatments), pre-determined times of day of an observation, or clearly categorised (e.g. age or sex) groups of animals. The behaviour or trait which is measured is called the **dependent variable**. Dependent variables are thus the response of the animals to an experimental stimulus or in the previously defined observation categories (times of day, age groups, sex). In other words, the independent variable is the *cause*, and the dependent variable is the *effect*. **Usually, data sets have several dependent and independent variables.** It is useful to consider the consequences of such data structures for statistical data analyses and interpretation (▶ Sect. 7.1). The terms "dependent variable" and "independent variable" are standard terms as used in many statistical programs and are key in coding the statistical analyses in free statistical software environments such as R (▶ https:// www.r-project.org) (R Core Team 2023; Jones et al. 2023). Note that dependent variables are also often called response variables or outcome variables, while independent variables may also be called predictor variables or explanatory variables.

Experiments are designed to manipulate one or a few influencing factors in a controlled manner to be able to determine causal relationships. However, an overly standardised environment can also negatively influence the reproducibility of behavioural data or limit the general validity of the results (Würbel 2000; Richter et al. 2010).

When designing an experiment, it is important to consider how to analyse the data. With a poorly structured data protocol or experiment, the subsequent analyses and can lead to biases from unplanned decisions during the analyses. The different available statistical methods have specific requirements for the data structure and the experimental design (Quinn and Keough 2002; Jones et al. 2023). For many statistical tests the data must meet certain requirements, e.g., be independent from each other. We also need to consider that we collect sufficient datapoints/replicates (▶ Sect. 3.2.2) for a statistical test to detect a possible biologically relevant effect. Likewise, in case the analysis needs to control for a complex nested test design, we should be aware beforehand which types of analyses will be required. **Power analyses** (▶ Box 3.7) carried out prior to a study, for instance, can help to estimate how large a sample must be in order to statistically verify an expected effect. If we expect very small effects, we can then determine in advance that a large sample size will be required to detect significant differences between responses to different stimuli or between experimental groups. Such a priori calculations of the required sample size are often required to justify the required number of animals when applying for permissions to conduct animal experiments. In general, we should always choose sample sizes carefully in order to find a reasonable balance between the power of the experiment and the number of animals being tested (also in line with the 3Rs principle; ▶ Chap. 2).

3.2.1 Controls

In experimental research, the control condition/group is crucial. Control conditions are necessary to determine whether an animal in an experimental context responds to a particular stimulus or, more generally, displays behaviour that is specific to a particular context. In some circumstances two or more different controls may even be necessary. It also is important for controls to be well planned and to be integrated into the timing of the experiment. If the control experiments are conducted after the actual experiments, timing is an additional factor, as different behaviour in the experimental and control conditions may be caused by external circumstances (different climatic conditions, different internal conditions of the animal, experience of the experimenter, etc.). Also, an observer may develop expectations about the behaviour in the control condition, may unconsciously be less careful during a longer experimental phase, or may introduce other biases over time (▶ Chap. 5). These and other sequencing effects can be counteracted by randomising or balancing the experimental conditions.

3.2.2 Data Recording Structure

One of the necessary decisions when you plan an experiment is the distribution of the test animals to the experimental test and the necessary control conditions. In simple experimental protocols there are two possibilities: 1) each animal is tested under both experimental and control conditions; 2) some animals are tested under one experimental condition while other animals are tested under the control condition (▶ Box 3.2). The same is true if several experimental groups are part of the study design. In the first case, this is called **paired sampling** (or repeated measures experiments) (▶ Box 3.2). If each individual animal is tested under only one condition, this is referred to as **unpaired, or independent, samples**. Specific statistical procedures are designed to address each of these data structures (▶ Box 3.2). Experimental designs in which some of the animals are tested under two or more conditions and some only under one test condition are more problematic to analyse and ideally are avoided. This is another example of how important it is to decide early in the experimental design stage which statistical tests should (or can) be used and which requirements these tests place on the data structure.

Box 3.2: Paired/Dependent and Unpaired/Independent Samples
The main differences between paired (◨ Fig. 3.2a) and independent samples (◨ Fig. 3.2b) are as follows:

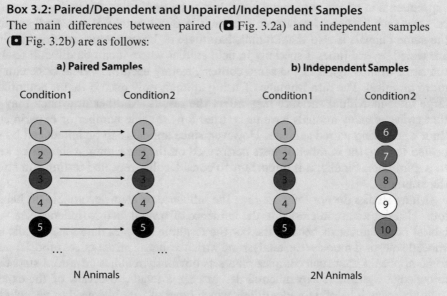

◨ **Fig. 3.2** Experimental designs for paired (**a**) and independent samples (**b**). In the case of paired samples, each animal (symbolised by the numbered coloured circles) receives both test conditions. For independent samples, each animal is assigned only one test condition

- In paired samples, each animal is tested repeatedly, in independent samples only once.
- In paired samples, strong individual differences between animals will be taken into account, while in independent samples, strong individual differences may override the effects of experimental conditions.

- In paired samples, one needs to consider sequence effects, as experience with the first test may affect the second test; this is not necessary with independent samples.
- For paired samples, the number of animals is independent of the number of test conditions, whereas for independent samples the number of animals required increases with the number of test conditions.

Both approaches presented in ▶ Box 3.2 have their advantages and disadvantages. The advantage of exposing each individual animal to all test conditions is that we can consider (and control for) in the data analysis the response of each animal in relation to both test conditions. Thus, we can control for individual variation in responses. For example, individuals which are very active or very inactive are compared with themselves under different experimental conditions. If we expect large variation in behaviour between different individuals, then experiments with such repeated measurements on the same animal are most suitable. Sometimes, however, repeated testing of the same individual is practically not possible or not appropriate to answer a research question. Invasive experiments or even learning experiments in which only naive animals can be used once, for instance, usually rule out such a repeated-measures approach. A disadvantage of repeatedly testing the same animals is also that animals habituate or become more sensitive if they are tested several times. Especially in field studies where it can be difficult to find the same animal again in the same context, testing each individual once can be more practical. The disadvantage of using animals only once is that a potentially large inter-individual variance may affect the results. Another drawback may be that twice as many animals have to be used for the same number of experiments than when using paired samples. However, since animals can be influenced by repeated testing, the number of tests performed on the same animal should be kept to a minimum. Finally, it is important to consider that repeated testing can stress the animal.

Many studies do not only examine the influence of a single variable on behaviour. Usually, we are interested in the influence of several factors (independent variables) on a number of behaviours. For the resulting complex data sets, specific advanced statistical models (e.g. analysis of variance, linear mixed-effect models) need to be applied. Using such complex data sets obviously requires advanced statistical knowledge. Systematic experimental designs, e.g. a tabular overview of the experimental design, is helpful for identifying which types of variables need to be included in the analysis and which types of analyses will be required. Classical examples of the more complex designs are randomized blocks or nested variable arrangements.

In a **randomised block design,** the experimental animals are randomly assigned to two or more experimental blocks. The term "block" is derived from examples in ecology (▶ Box 3.3). Here, one can imagine an experimental field in which a researcher tests three experimental conditions (e.g. different vegetation densities). Each of the three test conditions is then assigned to several small test fields (replicates). Since there may be environmental gradients in one direction of the test field (e.g. soil moisture gradients or shade), it is useful to divide the test field into blocks and to randomize the test conditions within these blocks. On the one hand, this allows the block to be included in the statistics as a factor or random factor, and on the other hand it ensures that the replicates are better distributed over the test field.

Box 3.3: Randomized Blocks Design

This results in a more uniform distribution of the experimental conditions over the total area (◘ Fig. 3.3) than in the case of randomisation without block formation. Likewise, the three test conditions (A–C) are more evenly distributed over differences in environmental conditions, such as differences in vegetation and soil moisture.

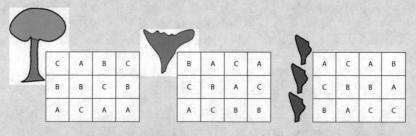

Gradient across soil moisture, differences in vegetation and light intensity

◘ **Fig. 3.3** A study field is first divided into blocks (the large fields) before the randomisation of the test groups (A–C). Within these blocks the experimental conditions are then randomised

More generally, the randomized blocks can also be presented as a table (◘ Fig. 3.4).

Experiment

Block 1 (Room 1 or Group 1)				Block 2 (Room 2 or Group 2)				Block 3 (Room 3 or Group 3)			
A	B	C	B	A	B	C	B	A	B	C	B
A	C	C	A	A	C	C	A	A	C	C	A
B	C	B	A	B	C	B	A	B	C	B	A

◘ **Fig. 3.4** The blocks can be, for example, fixed factors such as test rooms, animal groups, or times of day. Within the blocks the experimental conditions (A–C) are then randomly distributed among the twelve individuals in this example

3

More generally speaking, however, the blocks do not have to be spatial areas, but are any set of fixed factors controlled by the experimenter, such as times of day, groups of animals within which individual individuals are tested, or even experimental rooms which may differ, for example in microclimate. Within these blocks, the individuals are randomly assigned the different experimental conditions. This can be done either by rolling a dice or by using random numbers, e.g. in R (R Core Team 2023) (▶ Box 3.4).

A blocked or randomized experimental design makes it possible to include the test design in a statistical model (e.g. as random effects) and thus to control for a possible influence of experimental design on the behaviour of individuals.

Box 3.4: Random Assignment Option

Possibility of randomly assigning individuals to experimental conditions. Random numbers can be calculated in Excel or R, for example (◘ Fig. 3.5).

Step 1 Insert random numbers		Step 2 Order random numbers upwards		Step 3 Assignment of Treatments (A orB)		
AnimalID	**Random number**	**AnimalID**	**Random number**	**AnimalID**	**Random number**	**Treatment**
1	0,8753	2	0,3072	2	0,3072	A
2	0,3072	4	0,3401	4	0,3401	A
3	0,4692	3	0,4692	3	0,4692	B
4	0,3401	1	0,8753	1	0,8753	B

◘ **Fig. 3.5** One possible approach is to assign a random number to each animal (step 1) and to sort the individuals according to the random numbers in ascending order (step 2). In this way, the individuals are randomly mixed (randomised). Then the experimental conditions can be assigned to the blocks (step 3). The actual order of the experiments can then be determined in the same way

Examples of commands to generate random numbers in R

— *runif(n=*NUMBER, *min=*0, *max=*1*)*; the command generates random numbers under the assumption of an equal distribution of the values in the number range 0 to 1; however, this range can be freely determined by *min* and *max*. The number of random numbers to be acquired must be defined and inserted accordingly in the command at NUMBER. It is also possible to output random numbers for other distributions such as binomial distribution, normal distribution, etc. Link to free software R (R Core Team 2023): ▶ https://r-project.org

In the below shown screenshot of R, NUMBER=10 was used:

```
> runif(n = 10, min = 0, max = 1)
 [1] 0.78836958 0.96818364 0.10732760 0.06369696 0.48771169 0.63411617
 [7] 0.14693105 0.11972499 0.90946003 0.17132895
```

In a **hierarchical test design,** several test animals are more closely related to each other than to other test animals in a clustered manner. A classic example is the growth of young animals under different experimental conditions (e.g. Honarmand et al. 2010; Krause et al. 2017). If different families are tested separately, it must be taken into account that relatives or animals that grew up together do not represent completely independent data points. Here, siblings must be nested, e.g. within their mother's ID. In the data analyses, the hierarchical data structure makes it possible to mathematically separate the variation arising from sharing a mother from the variation between individuals (▸ Box 3.5) (Quinn and Keough 2002; Schielzeth and Nakagawa 2013; Jones et al. 2023).

Box 3.5: Hierarchical Test Designs
Hierarchical data structure using the example of broods of three mothers, which are either completely assigned to test condition A or B (◨ Fig. 3.6).

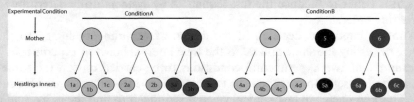

◨ **Fig. 3.6** A hierarchical data structure allows all data points to be included in the analysis. In this case, all 16 young animals can be considered without problems of pseudoreplication (▸ Sect. 3.5.2); the mother ID is considered as a nesting random factor, e.g., in a linear mixed effect model using the package 'nlmer' (Pinheiro et al. 2023) in R (R Core Team 2023). Without a nested experimental protocol and the correspondingly nested data analysis, the sample would only be N=6, since each condition affects all juveniles in the nest simultaneously and these data points would have to be averaged accordingly

---

3.2.3 Order Effects

Where different tests are carried out on the same animals, the order in which data are collected must be well planned. Particularly in experiments where animals are sequentially exposed to different stimuli, effects of the order of the sequence itself may affect their behaviour. **Animals can habituate or become more sensitive through repeated exposure to experimental treatments. Moreover, exposure to one treatment may affect the subsequent response to another.** To counteract this, the test sequence should be systematically varied (balanced) or randomised across the different test animals. This means, for example, that for 10 test animals, animals 1, 3, 5, 7 and 9 will first receive test condition A and then test condition B, while animals 2, 4, 6, 8 and 10 will each receive test condition B and then test condition A. We need to take similar precautions when each animal is assigned only one test condition. When animals are tested in a three-condition study, the animals should

3

first be randomly assigned to the test conditions (random assignment options; ► Box 3.4) and then the sequence of experiments should be randomised/balanced.

Balancing should not result in the experimental conditions always being systematically carried out at, for example, different times of day. If, in a balanced design, only two experiments can be carried out in one day, e.g. one in the morning and one in the afternoon, then the experimental conditions need to be balanced such that half of each test condition are in the morning and the other half in the afternoon, respectively.

Also, the expectations, concentration, or care of the experimenter may change with increasing duration of the experiment. Such an effect can be controlled by alternately performing the different test conditions, as it should then affect all test conditions in the same way. **For large samples,** test sequences are therefore usually randomised. **For small samples,** however, randomisation may not be very helpful, because it is possible that strong sequencing effects may occur by chance. In such cases, it is better to balance the test sequences by systematically determining the alternating order.

Another important decision in the design of experiments, linked to the sequence of testing the same animal, is the time interval between experiments. Take the example of two experimental conditions to be carried out on the same animal. It can be useful to carry out the two conditions in as close as possible succession. This approach can be practical when it is difficult to find the individual to be tested again in a similar context. The disadvantage, however, may be that the reaction to the first experimental condition persists beyond the direct experimental context and thus directly affects the next experimental condition. If there is a strong risk that the response from one test "spills over" to another, it is better to avoid such paired sample testing, especially in short succession. Independent samples, in which each individual animal is tested under only one test condition, are more appropriate in this case.

Box 3.6 Cautionary Example: Side Preferences in Choice Experiments

One classic, widespread experimental approach is to give animals a choice between different alternatives (e.g. Holveck and Riebel 2007; Caspers et al. 2009; Lea and Ryan 2015; Krause 2016; Gierszewski et al. 2017). Such experiments can be, for example, choice experiments in which an animal is exposed simultaneously with two different stimuli on different sides (◘ Fig. 3.7). In such or similar experiments we need to be aware that animals usually have side preferences that are independent of the experimental stimulus. Such side preferences may be due to some quality of the testing arena, for example that a test room is not symmetrical, is unevenly lit, or one side is adjacent to other animal housing spaces, conditions found in many laboratories. Furthermore, side preferences can be based on individual differences (e.g. laterality, such as right- and left-handedness in people) and have behavioural, physiological, or developmental causes. Therefore, side preferences can also develop in a very hidden way without being obvious to an experimenter.

☐ **Fig. 3.7** Possible experimental set-up for choice experiments, here using the example of a species identification experiment. In the example, a male diamond finch (*Stagonopleura guttata*) has the possibility to choose between a conspecific male (left) and a heterospecific male zebra finch (*Taeniopygia guttata*). In such an experimental set-up, the duration of presence in the preference zones (the zones in which it is considered as choice of one or the other stimulus; here the areas highlighted in blue) can be caused by side preferences independent of the experiment. Therefore, it is necessary to control for side preferences, e.g. by changing sides of the stimulus animals after half of the experiment time

Such side preferences are a well-known and frequently occurring phenomenon. Therefore, when two sides of a stimulus are presented (e.g. in simple partner preference experiments), the side from which a particular category of stimulus is presented should be changed for different animals (☐ Fig. 3.7). A direct way of recording side preferences as such is to divide a choice experiment into two phases and to change the sides of the stimulus between the phases (Witte and Sawka 2003).

3

3.3 **Sampling**

Sample size is a key consideration in any data collection. How large must a sample be in order to draw sound conclusions on a group, population, species, or a specific taxon when we study only a limited number of individuals? How large must a sample be if we only want to determine capabilities of an animal species, for example in terms of cognitive potential?

The quality of a study is by no means equal to the quantity of the sample. In this sense, we need to carefully estimate which sample size is possible and adequate. Such an estimation is particularly important in the case of animal experiments that are subject to authorisation, for which the sample size must be justified and may be limited for legal and ethical reasons (Still 1982; Ruxton 1998). In any case, a priori calculation or estimation, if possible, of your sample size is useful. Sample size is generally more limited in studies involving vertebrates than in studies involving invertebrates, which can be kept in large numbers in a small area and are often found in large numbers in the field (▶ Box 3.7). In studies where we plan to analyse the data using simple statistical tests, small samples may be sufficient if the effects are very strong, e.g. if all animals in one test condition behave systematically differently from those in another test condition.

Box 3.7: Sample Size
The significance of a study to be generalized generally increases with increasing sample size (◘ Fig. 3.8). However, practical and ethical considerations must also be considered when determining a sample size. Above a certain level, an increase in

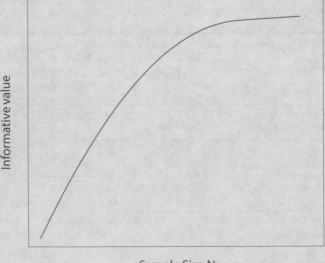

◘ **Fig. 3.8** Relationship between sample size and significance (external validity) of a study. The extent and relevance of this correlation depends on the research question and other factors

the sample size does not lead to an additional significant gain in knowledge. An increase in the sample size must not lead to compensating for the additional workload by less careful data collection on the individual animals.

Appropriate sample sizes (N) can be calculated in advance with various software tools, e.g. with the free G*Power 3 (Faul et al. 2007), and thus determine which N is required to detect a biologically relevant effect of a certain strength at all. (Link to the G*Power: ▶ http://www.gpower.hhu.de/).

Therefore, it is important that we try to estimate in advance the minimum number of animals that we need for a statistical test to be able to show significant differences (▶ Box 3.7). The smaller the expected effects are, the larger the sample needs to be. Statistical power analyses make it possible to estimate whether a sufficiently large sample can be achieved for a specific problem with a reasonable amount of effort. As a rule, when there is strong individual variation, i.e. different individuals behave very differently in the same situation (▶ Box 3.8), then the statistical minimum sample will not be sufficient. It is also often necessary and desirable to analyse the data collected using multivariate statistics. As a rule of thumb: The more factors to be considered in the analysis, the larger the sample should be. Pilot studies and comparisons from the literature, on the basis of which we may be able to estimate the expected effects, are certainly helpful in planning the sample.

It is not acceptable to start analysing the data during data collection and to stop a study at the precise moment when the data have just become significant. Such an approach is against good scientific practice. Therefore, while designing an experiment or any data collection, it is important that we consider which sample size is sufficient and at which point the data collection will be completed, which is usually when the targeted sample size is achieved or when the time planned to collect the data is over.

3.4 Individual Differences

Individuals of a species often differ considerably in a variety of behaviourally relevant traits (Wolf and Weissing 2012). Thus, there may be different ways (often related to the personality traits of an animal) to solve the same problem without one strategy having to be fundamentally better than the other. Such individual differences are particularly pronounced in more highly developed vertebrates, where development, experience and learning play a crucial role in behaviour (▶ Box 3.8). Individual differences are caused both by genetic and developmental factors (Naguib et al. 2011; Monaghan 2008; Briga et al. 2017) and by the circumstances currently affecting an animal. In behavioural studies in which we want to determine the significance of a stimulus, such behavioural differences could be disturbing if the variation in behaviour among animals exposed to a stimulus is greater than the variation between two groups of animals presented with different stimuli. For example, some individuals may respond strongly to a

3

stimulus by quickly approaching very closely but not showing any other specific behaviour (▶ Box 3.8). Other individuals may approach less closely but signal strongly from a distance (vocalisations, high activity), or overall just respond at lower intensities in general. These diverse behavioural strategies could reflect similar arousal, but these very different responses arise because individuals just express their arousal differently. This can be seen as similar to the way humans can have very different reactions to situations depending on if they are shy or bold. If there are several such different strategies for responding intensively to a stimulus, it is possible that no measured behavioural pattern will have significant effects on a particular stimulus, even though all individuals show strong responses. In such a case we want to use statistical analyses which consider this complexity. **Moreover, if the variation between animals is expected to be very large, a larger sample is all the more necessary.** We also need to be aware that even a clear standardization of the conditions for data collection still can have divergent effects on animals.

One elegant way to consider such individual- or personality—differences, is to quantify them, by for instance exposing individuals to so-called personality tests (van Oers and Naguib 2013; Carter et al. 2013). These are often open field (e.g. Walsh and Cummins 1976) or novel object tests in which we can quantify variation of behaviour in a context independent of the context to be studied subsequently (e.g. Takola et al. 2021). In order to consider such individual differences as characteristic for an animal, to reflect personality traits, these should be repeatable and/or have predictive value for behaviour in other contexts. A bold individual can be expected to be bold across contexts, while a shy individual is expected to be systematically shy. Such tests are best done when animals are under slight stress, such as in a novel non-dangerous situation, as then differences emerge more strongly than in contexts with no stress or extreme stress.

Since animals often behave differently depending on sex, age, social position, previous experience, and nutritional state, it is useful to take such factors into account when selecting study subjects (▶ Box 3.8). Individual variation is part of the behavioural repertoire of an animal species and can provide deep insights into the behavioural spectrum and reaction norms (Groothuis and Trillmich 2011). Therefore, when trying to minimize individual variation, one must not go so far as to make the conditions under which the animals are studied so abstract and unnatural that regular variation becomes lost. In such a case it becomes problematic to extrapolate the findings from the data recording to natural situations. For example, the standardisation of housing conditions to a low- standard or to unnatural social structures can lead to the animals varying little in their individual behaviour, though they would in more natural conditions. These findings are then only generalisable to a limited extent.

Box 3.8: Individual Differences
See ◘ Fig. 3.9.

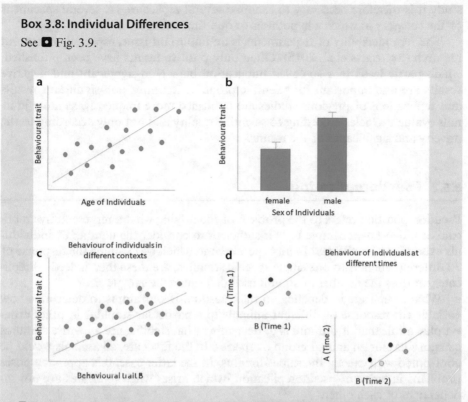

◘ **Fig. 3.9** Variation in behaviour between individuals can have a variety of causes, for example differences in age/experience (**a**) or sex differences (**b**) But even in individuals of the same age and sex, for example, there is variation in behaviour. Many animal species exhibit behavioural traits that are correlated over contexts (**c**) and time (**d**), which are referred to as animal personality (Wolf and Weissing 2012)

3.5 Replication and Pseudoreplication

3.5.1 Replication

Replications are necessary if we want to generalize the results of a study. Depending on the research question, replications can be, for example, the number of animals tested or the number of groups, trials or populations tested (► Box 3.9). If we study only one animal, it is likely that we collect data that are not well representative of a group, population, or species. Similarly, the behaviour of a single group need not be representative of a population or species. For a representative study it is therefore necessary to examine several replicates, i.e. several specimens of the category to which a hypothesis or question refers.

The reproducibility of experiments is an important issue, especially in current research (e.g. Aarts et al. 2015). Often only positive results have been published, which partly leads to a so-called publication bias. Non-significant and negative results are also important for scientific progress. Recently, non-significant results and replications of previous studies are published more frequently, as more journals evaluate articles according to scientific quality and not only according to the novelty and significance of the results.

3.5.2 Pseudoreplication

Pseudoreplication refers to the problem of identifying what a representative replicate or independent sample is. Is it sufficient to consider the number of individuals examined as replicates? In an experiment in which we examine the response of 30 different animals to one single specific stimulus, are these then indeed 30 replicates, or does the number of stimuli used determine the sample size?

What a replicate is, depends on the question. If our aim is to determine how variable the response of different animals to a particular stimulus is, other criteria play a role than if our aim is to determine what significance an entire stimulus category has for an animal group or species. In the first case, all animals would be confronted with exactly the same stimulus. In the latter case, this approach poses problems in terms of pseudoreplication, which arises when replicates are not independent of each other.

In practice, pseudoreplication occurs when categories are studied which vary in themselves, but only one stimulus is selected from a category and this stimulus is then repeatedly tested on several animals (▶ Box 3.9). Even if we study 20 animals using such a procedure, but we confront them all with the exact same stimulus, the final sample size at stimulus level still remains $N = 1$.

If pseudoreplication is present, a very elaborate and otherwise well-designed study may not be publishable or only with great restrictions, as its general scientific value is limited. Since it is usually difficult to demonstrate quantitatively that a single stimulus is actually representative of a category, we need to use several stimuli exemplars which then can better represent the existing variation within the category. This problem is biologically not trivial. If only one stimulus per category is selected, it cannot be ruled out that a particularly unattractive stimulus has been randomly selected in one category and a particularly attractive stimulus in another. Possible differences in the response to the two stimuli may not result from differences in the category at all, but from differences independent of the category (▶ Box 3.8).

Box 3.9: Different Types of Replications (◘ Fig. 3.10)

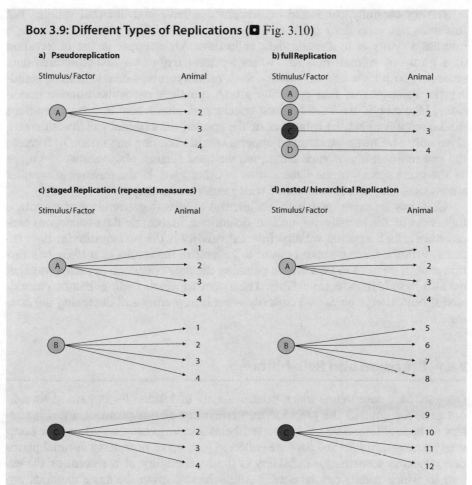

◘ Fig. 3.10 **a)** In pseudoreplication, one stimulus is used several times, so that the sample of stimuli remains small despite an increasing number of animals. **b)** In full replication, each animal receives a different stimulus of one stimulus category. Replicates of the stimuli and also of animals are used. **c)** in the case of grouped replication (with repeated measurements), several stimuli are used as replicates, but each stimulus is used several times with the same animals. The repeated use of the same stimulus is statistically accounted for in the analysis (e.g. as a random factor). **d)** Nested/hierarchical replication (▶ Box 3.5)

3.6 Internal and External Validity

Validity is the extent to which the project accurately measures what it is supposed to measure. Scientific studies can be distinguished in their so-called internal and external validity. **Internal validity** is the significance of a study in relation to the concrete situation that is being studied. **External validity** is the degree to which the findings from a study can be generalised beyond the example that is investigated.

3

A very carefully conducted experiment can have high internal validity, but this does not necessarily relate to its external validity. The prerequisite for high internal validity is systematic data collection. An example is the observation of a group of animals in a zoo, where we can carry out a well-structured data recording with high internal validity. Such results can be valuable in understanding the behaviour of that particular group and their particular housing conditions. However, if we want to draw conclusions about how housing conditions fundamentally affect the behaviour of the species, the external validity is limited, since only one group was studied under a specific housing condition. To increase the external validity of such a data set, we need further observations of groups of the same species in the same zoo or in other zoos. If the findings are similar across sites, they can be interpreted with greater external validity.

Certainly, an experiment with low internal validity (e.g. from lack of a control, influences of the investigator, unclear definitions, inaccurate data collection) cannot have a high external validity. Internal validity is the prerequisite for high external validity. Since we usually want to generalize the results of a study to a certain extent, we need to take care in planning the data collection in such a way that we also have high external validity. The extent to which findings can be generalized should always be viewed critically when interpreting and discussing the findings.

3.6.1 Precision and Reliability

Once we have determined which behaviours are of interest for our study, we cannot take for granted the precision and reliability of our protocol. **Precision** refers to how differentiated the observed behaviours are categorized or how accurately metric or numerical data are collected (e.g. up to how many decimal places one measures something). **Reliability** is the repeatability of a record, or the extent to which results can be consistently achieved using the same methods under the same circumstances. A particularly reliable observation need not be extremely precise, in the sense that a second observer comes to exactly the same result. However, reliable measures do not ensure validity, as two observers may erroneously record the same result, for example if the data collection is influenced by their expectations (▶ Chap. 5). The problem of imprecise definitions of behaviours or categories of behaviour also becomes apparent when we repeatedly log the same sequence from a video recording and come to a different result from one run to the next. In this respect, it is important to choose behavioural categorisations and measurement accuracy in such a way that repeated logging of the same behavioural sequence produces the same result (▶ Chap. 5). If the two outcomes are largely comparable, the observation methodology is said to be highly reliable (Caro et al. 1979; Jones et al. 2001).

References

Aarts AA, Anderson JE, Anderson CJ, Attridge PR, Attwood A, et al (2015) Estimating the reproducibility of psychological science. Science 349:aac4716

Briga M, Koetsier E, Boonekamp JJ, Jimeno B, Verhulst S (2017) Food availability affects adult survival trajectories depending on early developmental conditions. Proc R Soc B 284:20162287

Caro TM, Roper R, Young M, Dank GR (1979) Inter-observer reliability. Behaviour 69:303–315

Caspers BA, Schroeder FC, Franke S, Streich WJ, Voigt CC (2009) Odour-based species recognition in two sympatric species of sac-winged bats (*Saccopteryx bilineata*, S. leptura): combining chemical analyses, behavioural observations and odour preference tests. Behav Ecol Sociobiol 63:741–749

Carter AJ, Feeney WE, Marshall HH, Cowlishaw G, Heinsohn R (2013) Animal personality: what are behavioural ecologists measuring? Biol Rev 88:465–475

Chakarov N, Pauli M, Krüger O (2017) Immune responses link parasite genetic diversity, prevalence and plumage morphs in common buzzards. Evol Ecol 31:51–62

Faul F, Erdfelder E, Lang AG, Buchner A (2007) G*Power 3: A flexible statistical power analysis program for the social, behavioral, and biomedical sciences. Behav Res Methods 39:175–191

Gierszewski S, Müller K, Smielik I, Hütwohl JM, Kuhnert KD, Witte K (2017) The virtual lover: variable and easily guided 3D fish animations as an innovative tool in mate-choice experiments with sailfin mollies-II. Validation. Curr Zool 63:65–74

Groothuis TGG, Trillmich F (2011) Unfolding personalities: the importance of studying ontogeny. Dev Psychobiol 53:641–655

Holveck MJ, Riebel K (2007) Preferred songs predict preferred males: consistency and repeatability of zebra finch females across three test contexts. Anim Behav 74:297–309

Honarmand M, Goymann W, Naguib M (2010) Stressful dieting: nutritional conditions but not compensatory growth elevate corticosterone levels in zebra finch nestlings and fledglings. PLoS ONE 5:e12930

Jones AE, Ten Cate C, Bijleveld CJH (2001) The interobserver reliability of scoring sonagrams by eye: a study on methods, illustrated on zebra finch songs. Anim Behav 62:791–801

Jones E, Harden S, Crawley MJ (2023) The R book, 3rd edn. Wiley, Hoboken

Kerr NL (1998) HARKing: hypothesizing after the results are known. Pers Soc Psychol Rev 2:196–217

Krause ET (2016) Colour cues that are not directly attached to the body of males do not influence the mate choice of zebra finches. PLoS ONE 11:e0167674

Krause ET, Krüger O, Schielzeth H (2017) Long-term effects of early nutrition and environmental matching on developmental and personality traits in zebra finches. Anim Behav 128:103–115

Krüger O, Lindström J (2001) Lifetime reproductive success in common buzzard, *Buteo buteo*: from individual variation to population demography. Oikos 93:260–273

Lea AM, Ryan MJ (2015) Irrationality in mate choice revealed by tungara frogs. Science 349:964–966

Loning H, Griffith SC, Naguib M (2022) Zebra finch song is a very short-range signal in the wild: evidence from an integrated approach. Behav Ecol 33:37–46

Monaghan P (2008) Early growth conditions, phenotypic development and environmental change. Philos Trans R Soc B 363:1635–1645

Naguib M, Floercke C, van Oers K (2011) Effects of social conditions during early development on stress response and personality traits in great tits (*Parus major*). Dev Psychobiol 53:592–600

van Oers K, Naguib M (2013) Avian personality. In: Carere C, Maestripieri D (Hrsg) Animal personalities: Behavior, physiology, and evolution. The University of Chicago Press, Chicago, pp 66–95

Pinheiro J, Bates D, R Core Team (2023) nlme: linear and nonlinear mixed effects models. R package version 3.1–162, <▶ https://CRAN.R-project.org/package=nlme>

Quinn GP, Keough MJ (2002) Experimental design and data analysis for biologists. Cambridge University Press, Cambridge

R Core Team (2023) R: a language and environment for statistical computing. R Foundation for Statistical Computing, Vienna. ▶ https://www.R-project.org/

Richter SH, Garner JP, Auer C, Kunert J, Würbel H (2010) Systematic variation improves reproducibility of animal experiments. Nat Methods 7:167–168

3

Ruxton GD (1998) Experimental design: minimizing suffering may not always mean minimizing the number of subjects. Anim Behav 56:511–512

Schielzeth H, Nakagawa S (2013) Nested by design: model fitting and interpretation in a mixed model era. Methods Ecol Evol 4:14–24

Still AW (1982) On the number of subjects used in animal behaviour experiments. Anim Behav 30:873–880

Takola E, Krause ET, Müller C, Schielzeth H (2021) Novelty at second glance: a critical appraisal of the novel object paradigm based on meta-analysis. Anim Behav 180:123–142

Walsh RN, Cummins RA (1976) The open-field test: a critical review. Psychol Bull 83:482–504

Witte K, Sawka N (2003) Sexual imprinting on a novel trait in the dimorphic zebra finch: sexes differ. Anim Behav 65:195–203

Wolf M, Weissing FJ (2012) Animal personalities: consequences for ecology and evolution. Trends Ecol Evol 27:452–461

Würbel H (2000) Behaviour and the standardization fallacy. Nat Genet 26:263

Quantification of Behavioural Processes

Contents

Behaviour must be quantitatively assessed to answer scientific questions through statistical analysis. We can count or measure some types of behaviour relatively easily, but often behaviours are not easy to define or to distinguish from one another. In these cases, we need to establish clear and comprehensible definitions. For complex behavioural processes, preliminary observations can help to develop definitions that can be practically applied and which are meaningful with respect to the research question.

A structured approach for quantifying behavioural processes includes:
— Decide which behaviours are relevant for your question
— Conduct preliminary observations of those behaviours
— Make appropriate definitions for the behaviours (e.g. an ethogram)
— Define the data collection protocol (see sampling methods)
— Conduct a pilot study to test if the definitions and protocol work in practice
— Write down the methods and their justification
— Conduct the study

4.1 Selection and Definition of Behaviour

Which behaviours are important to answer a biological question? This simple question is not always easy to answer. To do so, we require experience and knowledge of the study species as well as the relevant literature on the topic. Depending on the animal species and the research question, different behaviours and recording methods are appropriate. Once we have identified the relevant behaviours, we then need to define them clearly enough so that someone else can reliably apply this definition. Behavioural definitions should not be subject to interpretation by the observer and therefore should avoid containing any subjective descriptors (see ▶ Box 3.1), as well as, in most cases, avoid making assumptions about the function of the behaviour. For example, you may be inclined to call a peacock opening his train ('tail') a "courtship display" but it is better termed "train fanning display" or "train rattling". Courtship would be an assumption of the function of the behaviour and should be avoided, as some behaviours serve multiple functions (e.g. a peacock often opens his train for courtship but may also do so to intimidate predators (Jordania 2021) or competitors (Paranjpe et al. 2022)). Moreover, very similar behaviours can serve different functions for different species. For example, in animal personality research the exact same behaviour (e.g. number of unique areas visited) in the same test (e.g. open field test) can mean something different (Carter et al. 2013), because individuals of some species feel safe (e.g. they have an overview) while those of other species perceive danger (e.g. they feel exposed). Labelling such behaviours as 'boldness' or 'exploration' rather than 'number of unique areas visited' can lead to misinterpretations of the personality, yet such terms often become standard terminology. Finally, we should aim to define behaviours in such a way that they are mutually exclusive. Even behaviour that appears very simple can gradually "fade in" and "fade away", so that

◘ Fig. 4.1 Brent geese during feeding and vigilance behaviour. The vigilance behaviour (defined as "head up") is in principle easy to count. However, even here there are gradual transitions, e.g. when the head is only half lifted, so that discrete boundaries have to be defined.

it must be clearly defined at which threshold the behaviour is recorded as such (◘ Fig. 4.1).

More complex behavioural sequences consisting of different elements, e.g. in complex displays are often less easy to define in detail. Thus, in such cases it is particularly important to determine in advance, through preliminary observations or literature studies, which elements of the behaviour may be important for answering the question. It is usually not possible to quantify individual components in a direct observation of a fast-moving animal. The use of audio recorders, video, or sensors, with subsequent detailed analysis, is a better alternative (▶ Chap. 6).

4.2 Categorizations of Behaviour

It is sometimes useful to group different types of behaviour into higher-level categories (e.g. 'foraging'), depending on your research question and on whether the categorization involves making any assumptions (see above). If, for example, you are interested in the foraging behaviour of bumble bees, distinctions could be made between the flight phase to a flower, the time at the flower, the time of food intake, and the flight time to the next flower. Recording of the individual components then allows you to derive decision rules for the search for food and to develop mathematical models regarding if and how the foraging is optimized. However, if your research question is focussed on the total time spent foraging compared to other behaviours, such detailed recording of the different behavioural components is not always necessary. Nevertheless, a researcher who is interested in a broad question, like 'time spent foraging', may still decide to collect the data as detailed as possible. The advantage of detailed recording is that it keeps the options open to do a more thorough explorative analysis than initially planned. Remember, behavioural components can always be grouped back into a category, but the other way around is not possible.

Usually, decisions about the level of categorisation of behaviours depend not only on the research question but also on the feasibility of recording behaviours at a fine resolution. Detection constraints can be a problem for collecting very detailed categories of behaviour, for example if behaviours occur very quickly or if animals are observed at a distance. Moreover, it is often the case that more detailed categories of behaviour take a longer time to record, for example if extracting them from video recordings, and you may simply not have this time to do this.

4   **As with defining behaviours for data collection, it is best if the level of categorisation is planned already before data collection begins.** These *a priori* decisions reduce the risk of bias. For example, it prevents the possibility that after the study behavioural components are unconsciously combined in a such way that they precisely lead to the expected results (see ► Chap. 5).

4.3   Defining Behavioural Bouts

Behaviour does not occur in random sequences. Some behaviours occur more frequently in a short period but then are not observed for a longer interval of time. Such a temporally periodic occurrence of the same behaviour is defined as a 'behavioural bout'. Examples can be the activity of an animal, its foraging behaviour, various social behaviours, or vocalisations such as bird singing bouts. The definition of such bouts commonly are biologically very important. But how can you decide when a behaviour is part of the same bout or already part of the next bout? There is no simple answer to this. **To define behavioural bouts we can best rely on statistical parameters as a criterion.** In principle, a wide variety of statistical measures can be considered, whereby those that depend on the distribution of the data collected are the most suitable (Fagen and Young 1978; Slater and Lester 1982; Sibly et al. 1990).

4.4   Time Parameters of Behaviour

Once we have defined the behaviours, we must determine how to record them. Knowing which time parameters you would like to extract from your observational data helps in deciding how to record particular behaviours (► Box 4.1). For example, when you are just interested in the latency or the frequency of a behaviour, you need to document only the start time of the observation and the start times of the behaviour. However, if you are interested in the duration, you additionally would need to document the end time of the behaviours. We discuss these different types of time-measures in more detail below.

Box 4.1
Decide before your observations which time-parameters you would like to extract
from the observed behaviour (■ Fig. 4.2):
- Latencies
- Durations
- Intervals
- Rates
- Frequencies
- Pauses

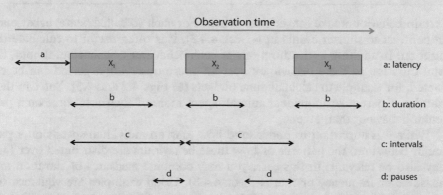

■ **Fig. 4.2** Time parameter of a behaviour (x). The different time characteristics can all be
relevant. Before starting a study, one should make predictions as to which time parameters
will change. Accordingly, not all parameters in a data collection are always of equal relevance

4.4.1 **Latencies**

Latency is often very important to measure, especially in experimental animal behaviour studies. The latency is the time it takes for an individual to respond to
a stimulus or to perform a particular behaviour (■ Fig. 4.2). Latencies are thus
powerful measures of how relevant a stimulus is for the animal. To measure latencies, one needs a starting and an ending point, which in experimental research
usually is the beginning of the experiment until the animal performs a specific behaviour.

However, an animal may not notice a stimulus immediately at the start of an
observation or experiment. In such situations it is better to define the latency as
the time from the apparent detection of the stimulus until the response. Likewise, if an animal is placed in a new environment and we want to determine how
long it takes to start interacting with an object, we need to make a distinction between the time from the start of the experiment and the time after the animal
has actually discovered the object. An animal that actively explores a test room
is likely to find an object more quickly than an animal that is initially quiet in a

corner. If we are not interested in the time an animal needs to find the object but rather the response to the object once it is detected, then we need to start measuring the latency from the moment the animal has detected it until it responds (e.g. approaches it). Latencies need not necessarily be measured directly as the precise number of seconds or minutes. In the case of interval-structured recording (▶ Sect. 4.5.2) rather than continuous recording, latencies can also be determined in the form of intervals.

4

4.4.2 Durations

Certain behaviours take longer than others. For such so called **'state' behaviours**, (in contrast to shorter events in ▶ Sect. **4.4.3**), it is often useful to calculate the duration. In addition to the durations of each behaviour or behavioural bout, the total duration of these behaviours over an entire observation period can be extracted, for example to calculate time budgets (◘ Figs. 4.2 and 4.3). You can then examine if particular individual animals spent more of their total time on a particular behaviour than others.

While it is important to understand how long an individual spends on a particular behaviour, the patterns of how these behaviours are distributed over time may also be relevant. In this case, you may combine measures of duration and measures of frequency or rate (▶ Sect. 4.4.3). Good examples are vigilance behaviour and foraging behaviour. Many short bouts of vigilance are more likely to reduce predation risk than one long vigilance bout and subsequent long periods without scanning the environment. Similarly, while the overall time spent foraging will determine the energy and nutrient intake, the duration of each foraging bout is relevant to better understand animal decision-making. For example, knowing how long animals stay on a particular patch before moving to the next is important when testing if animals follow an optimal foraging strategy.

Pauses or breaks between behavioural states (Fig. 4.2) can also be functionally very important, thus calculating the duration between bouts of behaviours can be valuable. For example, in communication, silences can have a very high signal value, so that quantifying their duration can be quite relevant for some questions.

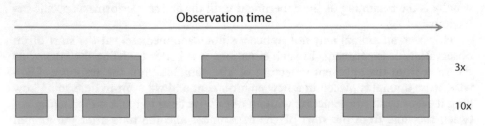

◘ **Fig. 4.3** Duration, frequencies, and rates of behaviour. The occurrence of the behaviours is symbolised by the boxes. Both types of behaviour have the same total duration, but occur with different frequencies and different individual durations. The data collection structure (▶ Sect. 4.5) should be adapted accordingly

Another example is rest periods between two energetically demanding behaviours, so the duration of the pause after a behaviour may be a good measure to determine the effort the animal has made.

4.4.3 Frequencies and Rates

If behaviours are very short, occur very frequently, or their duration (as in ► Sect. 4.4.2) is difficult to determine, it may be more useful to record only the frequencies, i.e. count how often they occur, and ignore the duration (◘ Fig. 4.3). Such behaviours are often called 'events' or 'point events'. When combined with the total observation time, we can also use frequencies to determine rates, i.e. events per unit time.

Rates are the number of occurrences per time unit, e.g. number per minute (◘ Fig. 4.4). Calculating rates is often the best way to standardize observations when observation periods differ across contexts or individuals. For example, if an individual was out of sight for part of the observation, we will have a shorter observation time than for those that we can see through our observation period. By calculating rates per time unit, results from different individuals, contexts or moments of observation can be standardised. A nice example for this are rates of vocalisations. In male songbirds, for example, the rate of vocalisations (number of songs/minute) during territorial defence or in response to females is important. Males that sing at higher rates may be in better condition than males that sing at lower rates (Snijders et al. 2017; Ritschard and Brumm 2012). It may take greater physical control to perform a behaviour very often in a short amount of time, compared to the same frequency of the behaviour but spread out over a longer duration of time.

The time for which to calculate the rate is not always straight forward. For example, if a behaviour does not occur in the observation period for a long time (the bird does not sing in 9 out of 10 observed minutes), then a calculated rate

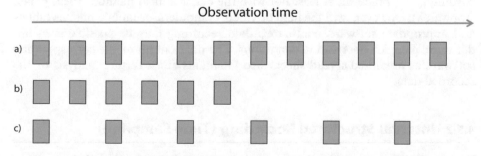

◘ **Fig. 4.4** When calculating rates, the distribution over time plays an important role. Here we assume a total observation time of 30 s. In **a)**, the rate is twice as large (12x/30 s=0.4) as in **b)** (6x/30 s=0.2) and **c)** (6x/30 s=0.2). Also note that the rate in b and c would be identical. If only the time in which the behaviour is shown is considered, the rate a) (12x/30 s=0.4) and b (6x/15 s=0.4) would be identical, but the rates between b) and c) (3x/30 s=0.2) would differ

over the 10 min would result in a very low singing rate and would not reflect the rate at which the bird sings when it actually sings (■ Fig. 4.4). It is therefore necessary to decide on the basis of the available data and the actual research question, which time window to calculate rates for and whether or not these rates adequately reflect the actual behaviour of the animals. Rates are most meaningful when the occurrence of the behaviour is approximately evenly distributed over time.

4

4.5 Recording Rules

Behaviour is a continuous process. However, it can be impractical or impossible to record behaviour continuously. Thus, in practice, various time-structured procedures are available, and these are elegant ways to capture the occurrence and time dynamics of behaviours (Altmann 1974; Tyler 1979; Naguib et al. 2013). In situations where we are not sure if different methods will yield similar results, it is best, of course, to test the different methods in preliminary experiments. On the basis of the results, we can then make an informed assessment of which recording rules are most suitable.

4.5.1 Continuous Data Recording

During continuous data recording, we record the beginning and end of the behaviour, so our data will accurately reflect the behavioural temporal dynamics. Continuous data recording is generally the method that comes closest to describing reality. The prerequisite for using this method is that we can define the behaviours precisely, i.e. determine the start and end points. For behaviours that gradually fade in and out, it is usually difficult to delimit the beginning and end. Another practical problem with continuous data recording can arise when several behaviours and animals are recorded simultaneously. In this case, continuous data recording is not practical, at least not with the classical field methods (pen, paper, stopwatch). However, with the help of video, computers, or mobile phones/tablets and appropriate software, continuous data recording may be possible even under more complex observation conditions. The data output of the corresponding software as organised spread sheets also facilitates direct further analysis of the recorded data.

4.5.2 Interval Structured Recording (Time Sampling)

For practical reasons, interval-structured data recording can be much more suitable. In interval-structured recordings, the time axis is pre-divided into intervals of equal length before the start of the data collection. If we choose appropriate sampling intervals, the data of an interval-structured recording can be almost as detailed as those from a continuous recording (■ Fig. 4.5).

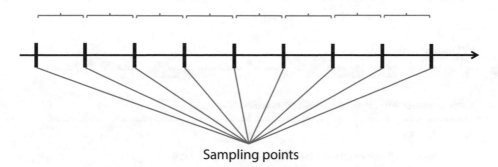

◘ Fig. 4.5   Interval structured time scale. With *one-zero sampling,* events that occurred within the previous interval are recorded at the sampling time points. With *instantaneous sampling,* events are recorded that occur at the exact sampling points

There are two common ways to collect time-structured data; one-zero sampling and instantaneous sampling. In **one-zero sampling,** we record at each pre-determined sampling point whether the behaviour had occurred (one) or had not occurred (zero) in the preceding interval. (**◘** Figs. 4.5 and 4.6). This method is particularly suitable when the behavioural characteristics are difficult to count. In **instantaneous sampling** (timepoint sampling), the behaviour is recorded only when it occurs at the exact sampling point.

Often we want to record several different behaviours, in which case it can make sense to combine different recording rules depending on the behaviour. Whether a behaviour should be recorded with *instantaneous sampling* or with *one-zero sampling* depends on the frequency of the behaviour as well as the duration of the behaviour relative to the interval duration (**◘** Fig. 4.6).

The general rule is that short-term behaviour (*events*) are better captured by *one-zero sampling*, whereas long-term behaviour (*states*) should rather be recorded with *instantaneous sampling* (◘ Fig. 4.7).

Instantaneous sampling is most suitable for behaviours that last longer than the duration of the sampling intervals, as otherwise all the instances when the behaviour occurs between the observation points would be missed. Very rare and/ or very short-lived behaviours such as a call, a gesture, or the picking of food are thus not suitable for recording in *instantaneous sampling*, but can sometimes be captured well with *one-zero sampling*. Yet *one-zero sampling* does not give a true frequency of the behaviour, but rather the number of intervals that contained the behaviour, and therefore can result in poor approximations of behavioural durations or frequencies. The strength of *one-zero sampling* is that it is usually easy to implement with a large number of behaviours and a large number of animals simultaneously. However, this benefit needs to be weighed against the potential for inaccuracy of the resulting data. In general, the shorter the intervals are the more

4

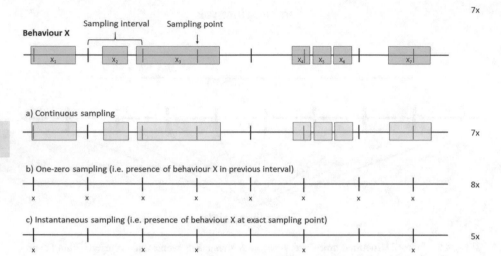

☐ **Fig. 4.6** Comparison of different registration methods. In addition to **a)** *continuous data recording*, depending on the duration of the behaviour relative to the interval duration, the different methods can map the temporal dynamics of the behaviour in different ways. With this interval structure, **b)** *one-zero sampling* leads to the behaviour being recorded as always present, i.e. 8x. The fine time structure would be lost. **c)** *Instantaneous sampling* is more sensitive to the fine time structure, but leads here to an underestimation of the behavioural frequencies (after Martin and Bateson 1993), i.e. 5x instead of 7x. Often the strength of interval registration lies in the combination of these two methods

accurate the data will be. When determining time budgets, for example, it should be kept in mind that data collected by interval-structured sampling are only useful if the interval durations are short enough to accurately approximate the relative allocation of the various activities (Pöysä 1991).

Although this structuring seems easy to implement at first, the choice of the time intervals cannot be solved ad hoc and tends to be much more complex in practice. Accordingly, it should be well thought through (and tested in a pilot phase), since the quality and significance of a study will depend on the choice of interval sizes (☐ Figs. 4.6 and 4.7). If the chosen intervals are too small, the quality of the data can suffer because there is too little time to accurately record all the behaviours before the next observation time point comes up, giving room for observation or recording errors to creep in. If the intervals are too large, important information is lost on the temporal dynamics and frequency of the behaviour. The size of the interval to be chosen depends on the behaviour to be recorded, the research question, and also on the skill and experience of the observer. So to stress again, **it is thus very important to test the appropriate duration of sampling intervals in a pilot study**.

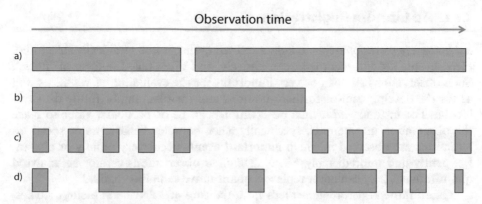

Fig. 4.7 Behaviours (orange bars) of different duration. Long-lasting behaviours (*states*); **a)** and **b)** are better recorded in *instantaneous sampling*. Short behaviours (*events*); **c)** and **d)** are better recorded in *one-zero sampling*

4.6 Sampling Rules

Next to different ways of recording behaviour, we can zoom out a bit and find that there are also different ways of sampling the behaviour (■ Fig. 4.8).

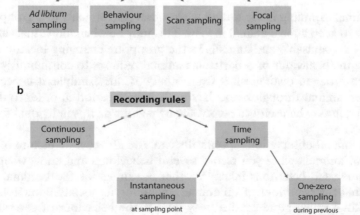

Fig. 4.8 An overview of the different **a)** sampling and **b)** recording rules available when conducting behavioural observations. Note that not all sampling rules combine well with all recording rules. For example, it will be near impossible to do continuous scan sampling, yet continuous sampling combines very well with focal sampling

4.6.1 Ad Libitum Registration

Ad libitum registration refers to a data recording that is not specifically structured. Here, events are not recorded quantitatively, and often take the form of 'notes'. Such observations still can be very important for the evaluation of results as well as for the development of further questions and the planning of future data collection. For instance, *ad libitum* observations can be of particular value to place the observation into context, specifically when complex behaviours or social interactions are observed in which important events occur very rapidly or also infrequently and unpredictably. Since *ad libitum* observations cannot be analysed quantitatively, they can never replace a quantitative sampling method.

Quantitative registration methods form the core of behavioural biology studies, which can only be supplemented, but not replaced, by *ad libitum observations.*

4.6.2 Scan Sampling and Behaviour Sampling

If groups of animals are observed, further decisions have to be made on how to record the behaviours. Is it better to record the behaviour of all animals at certain points in time or to focus on one animal at a time (focal animal sampling, see below)? When several animals are observed at the same time, **scan sampling** allows the behaviour of all animals to be recorded simultaneously at certain points in time (instantaneous sampling). Another possibility is to record the behaviour of interest in **behaviour sampling** whenever it occurs.

4.6.3 Focal Animal Sampling

Focal animal sampling is often used when observing animals in a group and it is necessary to keep a focal animal in view. This may be when individuals are not individually recognisable and animals in the group are changing locations, so that recording the behaviour of a particular animal requires to continuously follow it visually in order to distinguish it from others. If, for example, it is necessary to observe an animal through binoculars for a longer period in order to determine its identity, it may be useful to record the behaviour of this individual for as long as possible.

Focal animal observations specifically can give an accurate picture of the time budgets of animals, since you record several behaviours and know what percentage of time each behaviour takes. Another advantage of focal animal observations is that the behaviour of an animal, including its social interactions, can be recorded in a much more detailed way than if the behaviour of several animals is recorded at fixed time intervals. In a well-planned (random) sample of different focal animals, systematically varying over different times of day, comprehensive insights on the behaviour in larger groups can be gained. Drawbacks of this method lie mainly in the standardisation of data acquisition in relation to different focal animals observed at different times of day and in the different contexts. You thus need to have a good plan of balancing and randomly selecting your focal observations. Other challenges lie in avoiding following the same individual twice and in being able to follow one individual for long enough to actually collect sufficient data during an observation.

References

Altmann J (1974) Observational study of behavior: sampling methods. Behaviour 49:227–267

Carter AJ, Feeney WE, Marshall HH, Cowlishaw G, Heinsohn R (2013) Animal personality: what are behavioural ecologists measuring? Biol Rev 88:465–475

Fagen RM, Young DY (1978) Temporal patterns of behavior: durations intervals, latencies, and sequences. In: Colgan PW (ed) Quantitative ethology. Wiley, New York

Jordania J (2021) Can there be an alternative evolutionary reason behind the peacock's impressive train. Academia Letters: article 3534. ► https://doi.org/10.20935/AL3534

Martin P, Bateson P (1993) Measuring behaviour: an introductory guide, 2nd edn. Cambridge University Press, Cambridge

Naguib M, van Oers K, Braakhuis A, Griffioen M, de Goede P, Waas JR (2013) Noise annoys: effects of noise on breeding great tits depend on personality but not on noise characteristics. Anim Behav 85:949–956

Paranjpe DA, Mahimkar VR, Dange P (2022) Rethinking the functions of peacock's display and lek organisation in native populations of Indian Peafowl *Pavo cristatus*. bioRxiv ► https://doi.org/10.1101/2022.09.21.508866

Pöysä H (1991) Measuring time budgets with instantaneous sampling: a cautionary note. Anim Behav 42:317–318

Ritschard M, Brumm H (2012) Zebra finch song reflects current food availability. Ecol Evol 26:801–812

Sibly RM, Nott HMR, Fletcher DJ (1990) Splitting behaviour into bouts. Anim Behav 39:63–69

Snijders L, van Oers K, Naguib M (2017) Sex-specific responses to territorial intrusions in a communication network: evidence from radio-tagged great tits. Ecol Evol 7:918–927

Slater PJB, Lester NP (1982) Minimising errors in splitting behaviour into bouts. Behaviour 79:153–161

Tyler S (1979) Time sampling: a matter of convention. Anim Behav 27:801–810

Research Biases

Contents

© The Author(s) and Friedrich-Loeffler-Institut, under exclusive license to Springer-Verlag GmbH, DE, part of Springer Nature 2023
M. Naguib et al., *Methods in Animal Behaviour*,
https://doi.org/10.1007/978-3-662-67792-6_5

All scientific research faces potential biases that threaten to undermine the validity and/or reliability of its results, potentially leading to misinterpretation of the findings. Although they can rarely be entirely avoided or eliminated, awareness and detection of potential biases is key to minimising, or at the very least, identifying, their influence. There are many biases that are common across scientific fields of research, and there are a few that are specifically critical to animal behaviour research—namely, sampling biases and observer biases. These biases discussed here are distinct from deliberate false data recording (or the subsequent manipulation of data). Deliberate manipulation is of course prohibited according to the rules of good scientific practice. Biases, in contrast, are unintentional and/or subconscious for the researcher but could nevertheless cause false conclusions in a scientific study.

5.1 Sampling Biases

In animal behaviour research, our goal is often to extrapolate the behaviour of a subset of individuals to an entire population (or even species). Therefore, it is important to ensure that our subjects represent the greater population as closely as possible. When this does not occur, we can inadvertently end up with a sampling bias, or a non-random selection of study subjects (◘ Fig. 5.1). Such a sampling bias can result in research that is not replicable or generalisable, and thus lead to false general conclusions.

Sampling biases in animal behaviour research can arise through a number of factors. First, individuals themselves may have certain characteristics that make

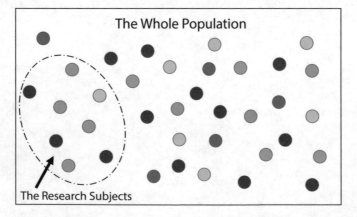

◘ **Fig. 5.1**   A sampling bias can occur if the research subjects do not accurately represent the greater population. The different coloured circles indicate subjects with different behavioural phenotypes with respect to the research question. Thus, a selection of research subjects as indicated by the dashed circle, that does not include all behavioural phenotypes and not in the proportion present in the greater population, leads to a sampling bias and results that are not representative of the population

them very different compared to the rest of the population (▶ Box 5.1). Another way that a sampling bias can occur is when subjects have already passed through a prior selection process, causing the researcher to focus on a subgroup of subjects that have survived this process and ignore those that have not. This type of sampling bias is specifically called **survivor bias**. For example, imagine a researcher wants to know how seriously a fall from a window is likely to injure a cat. To find the study sample, the researcher may visit an emergency veterinary clinic and record the injuries and associated window height of any cats that have been admitted after a fall from a window. Indeed, that is exactly the subject of a study from 1987 (Whitney and Mehlhaff 1987), which surprisingly found that cats that fell from windows less than six stories high had worse injuries than cats which fell from higher than six stories. The authors explained this seemingly miraculous result by proposing that the cats reached a terminal velocity after falling about five stories, after which the cats relaxed their bodies, minimizing injuries. Of course, an alternative explanation to this finding is that the cats who fell from greater heights *did not survive* to make it to the veterinary clinic, and therefore the findings were a statistical artefact, due to an overrepresentation of individuals that were lucky enough to not be killed by the highest falls.

Box 5.1 STRANGE Subjects

Researchers have identified a framework for identifying characteristics that can unwittingly cause a sampling bias in animal behaviour studies using the acronym STRANGE (Webster and Rutz 2020). STRANGE-related biases can affect both laboratory and field studies. They can have an influence on which animals are sampled during studies and the behaviours that they exhibit. Thus, care must be taken to evaluate study animals for these factors to minimize sampling biases and/or to discuss the potential influence they may have on study findings.

Social background can include any quality of an individual's past or current social status and social interactions, and any social learning experiences it may have had.

Trappability and self-selection can occur when individuals with certain traits (e.g. personalities) are more likely to be trapped or to voluntarily take part in experiments.

Rearing history comprises an individual's experiences during development, including its physical and social environment.

Acclimation and habituation occurs when study activities such as catching, handling, marking, or exposure to novel situations result in behavioural changes over time.

Natural changes in responsiveness can occur due to temporal (e.g. daily or seasonal) or reproductive cycles, or across different life stages.

Genetic make-up can generate distinct differences between individuals of wild populations or between wild and laboratory populations, and can have profound effects on behaviour.

Experience comprises past opportunities for learning, for example through participation in prior experiments.

Another example of survivor bias can be seen in studies of nesting success of wild birds. These studies rely on field researchers searching for and locating the nests that are to become the subjects of the study. However, active nests are generally much more likely to be found than those that have already failed due to predation or any other natural cause of mortality, or those where nestlings have fledged already. If you were to calculate the nesting success rate based only on the nests that have been found, you would potentially be missing a large sample of nests which were never found because they completed earlier or had already failed, thereby biasing the results in favour of current successful nests. Moreover, the farther along a nest is when it is discovered, the more likely it is to be successful, because it has already survived a period of risk which young nests have not yet experienced. Accordingly, only samples consisting of nests found on the first day they are established would yield unbiased success rates, and the bias increases with the age of nests at the time of discovery. However, including nests only found on the day they are initiated is just not realistic, and therefore older nests tend to be overrepresented in these studies, while nests that fail shortly after initiation tend to be underrepresented.

So how can a researcher account for samples that they do not have? One imperfect but sophisticated solution is to mathematically correct for the missing samples, which may not give an exactly accurate, unbiased answer, but which provides a closer estimate than ignoring earlier failed samples altogether. In the case of nesting birds, a method was developed to do just that (Mayfield 1961), variations of which have been implemented in countless studies since. In short, this method estimates a daily survival rate based on the number of nests that fail and the number of days monitored, summed across all observed nests. The daily survival rate can then be raised to the power of the duration (days) of the nesting period to calculate the probability of a nest surviving the full nesting cycle.

Other causes of sampling bias, such as the categories outlined in the STRANGE framework (▶ Box 5.1), are less straightforward to mitigate. Often, careful attention to study design (▶ Chap. 3) can reduce the influence of such factors, for instance by sampling across several populations and/or using multiple trapping techniques. For example, when in a lab population 10% of the animals should be tested, a selection of the first caught 10% could lead a bias to catch weaker or less attentive animals. A mitigation could be to not use the first 10% of the animals you have trapped/caught but instead catch all animals and then take every tenth into account. Another approach could be to list all individuals and then either randomly select 10% of the IDs, regardless of catching order, or directly try to catch these pre-selected individuals.

Any such measures taken to reduce sampling bias should be transparently reported, as should specific descriptions of any non-participating subjects. Findings should be explicitly linked to the studied sample rather than the greater population or species as a whole, in order to facilitate replication and comparison, and to inform the design of future studies. Through explicit identification of potential sampling biases, researchers can more adequately scrutinize methodology, and carefully consider and even debate the influence of such issues on the interpretation of the findings. A set of guidelines (ARRIVE guidelines (Animal Research:

Reporting of In Vivo Experiments), McGrath et al. 2010) has been established to help researchers to sufficiently report their live animal research in such a way to minimise biases and allow adequate scrutiny. It is recommended to consult these guidelines both during study design and during write-up.

5.2 Observer Biases

Data collection in animal behaviour research often relies on direct observation of behaviour. However, a risk with research based on observation is that the data can be directly influenced by the observer themself. Data collection must therefore be planned in such a way as to minimise or systematically account for any possible influence of observers. Since this is not always possible, all the more care must be taken to identify possible problems and take them into account when interpreting the results or to create control conditions in which the influence of an observer can be explicitly assessed.

Observer bias refers to any unconscious influence of an observer on the data. This can manifest in two different ways. On one hand, the presence of the experimenter can directly influence the behaviour of the subjects, also known as **experimenter bias** (Rosenthal and Fode 1963). On the other hand, observers can subconsciously influence data if, for example, they have a certain expectation of a result and they unwittingly record behaviour in such a way that supports that expectation (Nickerson 1998). This is called **confirmation** or **expectation bias**, whereby the behaviour of the animal is not influenced directly, but the results are rather influenced by the expectations of the observer. Finally, these two types of observer biases are not necessarily mutually exclusive. It can also happen that unconscious movements of the observer influence the animal's behaviour in such a way that it adapts its behaviour to the observer's expectations—a phenomenon also known as the **Clever Hans Effect**.

5.2.1 Experimenter Bias

In principle, an observer should position themself in such a way that they are noticed by the animals as little as possible. Observer hiding places may be set up in the field if it is necessary to observe animals at a specific location (e.g. a feeding station). Binoculars or spotting scopes are, in some situations, well suited to observe behaviour in detail even from a distance. In a laboratory, one-way panes can be useful, though because they act like mirrors from one side the position and lighting conditions must be set-up so that the animals do not see themselves. Alternatively, remote cameras can be used to observe the animals from an adjacent room or elsewhere.

Observers must also keep in mind that it may not only be their visual presence that is disturbing. Many animals can detect odours, as well as sounds, that we ourselves cannot perceive. Although the influence of an observer can be minimised if the behaviour is recorded via video or audio devices without the observer being near the animals, the placement and/or presence of such equipment can also influence behaviour. Researcher influence can also have delayed effects on data, for example the very act of a researcher checking a bird nest's status can directly influence the success of the nest by, for example, inducing nest desertion (Rodway et al. 1996), but can also indirectly influence nesting success when the researcher activity either deters (Lloyd and Plagányi 2002) or attracts (Strang 1980) nest predators to the nest sites. If an influence of the presence of an observer cannot be completely ruled out, the location of the observer should not vary between experimental groups, i.e. the position of the observer should be randomised or systematically varied (see ▶ Chap. 3).

5

5.2.2 Expectation Bias

Expectation bias arises from the human tendency to subconsciously search for, interpret, favour, and recall information in a way that confirms or supports our own beliefs or expectations (Nickerson 1998). In this way, observers can be inclined to connect ambiguous behaviours to their own expectations related to their research hypothesis, resulting in data that does not reflect differences in the true behaviour of the animals. For example, when university students were given the task of assessing panting behaviour of cows in videos (Tuyttens et al. 2014), they consistently rated the cows as panting more when a video was marked as having been taken in a higher ambient temperature, even when they were actually watching the same videos repeatedly (once with real temperature data and once with fictitiously elevated temperature data, unbeknownst to the students). In another test of this phenomenon, Brumm and colleagues (2017) asked volunteers to measure minimum frequencies in animal sounds recorded in noise and without noise. Half of these observers were provided with the hypothesis that animals vocalize higher in noisy conditions (Slabbekoorn and Peet 2003; Brumm and Naguib 2009). This "informed" group measured significantly higher frequencies only under the noisy conditions, but both groups came to the same measurements under the non-noisy conditions, demonstrating how the expectations of the observers can lead to false positive results. Moreover, observer effects can occur because researchers are often more inclined to re-measure extreme values that go in the "wrong direction", believing them to be erroneous. This reduces possible measurement inaccuracies in only one direction. Since these processes can occur unconsciously, it is important to be aware of this problem and to take appropriate precautions to minimize such influences.

   A first step in mitigating the influence of observers is to minimise the scope for subjective decision making when logging data by ensuring that boundaries between different categories of behaviour are clearly defined (▶ Chap. 4). Another way to reduce expectation bias is to leave the observer(s) "blind", by not informing them what the research question is or which experimental group the animals belong to, e.g. treatment and control group. In such experiments, however, it must be established that the naïve observers are able to collect the data with the same quality and systematic approach as an experimenter could. In many studies where behaviour is categorised directly at the time of observation, the experience of the observer plays a key role in reliably identifying the relevant characteristics. Therefore, in such special cases, it must be weighed whether the disadvantages of using naïve observers outweigh the risks of unconsciously influencing the data.

Finally, using multiple observers can facilitate detection of potential observer effects. If different observers record and/or categorize behaviour within an experiment, this usually leads to an increase in the variability of the measurements. Then, **the degree of agreement between the measurements of the different observers** can be calculated (e.g. Cohen 1968). If different observers arrive at the same result, it is considered high inter-observer reliability (Caro et al. 1979; Jones et al. 2001). This is usually a good indication that behaviours are objectively defined. However, note that multiple observers do not necessarily eliminate the problem of expectation bias. If several people are involved in the same project, a group effect can also occur, where all observers similarly record things in the way they expect. In fact, in both examples of expectation bias given above (Tuyttens ct al. 2014; Brumm et al. 2017), multiple observers yielded similarly biased results due to their shared expectations. Since researchers belonging to the same group are often testing the same hypothesis, care must be taken that there is not a group-wide bias influencing the research results.

5.2.3 Clever Hans Effect

A classic example of how an observer can unintentionally influence the behaviour of an animal to suit the hypothesis being tested is the amazing achievements of a horse from Berlin named Hans (Ferald 1983). Hans was apparently capable of solving complicated arithmetical problems, i.e. adding, subtracting, dividing, multiplying as well as root and fractional arithmetic. Hans delivered the answer to the presented problems by stamping the corresponding count on the ground with his right front hoof. With the correct answer, Hans was rewarded with a carrot. If, for example, Hans was asked how many of the men present were wearing straw hats, Hans answered with the correct number of kicks, not counting the number of women wearing straw hats. These curious and amazing feats were repeatedly shown at public performances and discussed on the front pages of the world press at the beginning of the nineteenth century (Fig. 5.2).

As a result of heated public discussions between convinced advocates of Hans' achievements and fierce critics that believed Hans' owner, a retired mathematics teacher, Wilhelm von Osten, was committing fraud, a high-profile commission was formed to investigate the phenomenon. The commission gave Hans various arithmetical tasks and specifically examined whether von Osten could have trained the horse's behaviour, for example by using hand signals when questioning the horse. However, it proved irrelevant whether von Osten was in front of or behind Hans, or whether another commissioner was asking the questions, even in

5

◘ Fig. 5.2   The "clever" horse Hans during a public interview

von Osten's absence. After intensive investigation, the commission concluded that there was no evidence of fraud and emphasised that von Osten was not signalling the horse.

In fact, the commission was correct that von Osten did not make any deliberate deception. He himself was convinced that it was exclusively Hans' skills and his teaching methods that led to these spectacular performances by the horse. However, following the commission's report, the psychologist Oscar Pfungst (1907) continued the investigation with careful scientific methods. After a series of experiments in which he systematically changed one variable at a time, Pfungst was able to show that Hans could only solve the tasks if he could see the questioner and if the questioner were aware of the solution to the task. Hans also solved the tasks less well the more distant the questioner was standing. Finally, it turned out that Hans also started to stomp his hoof when the experimenter bent slightly and ceased stomping when the experimenter bent back without ever having asked anything.

Pfungst's investigation was finally able to show that Wilhelm von Osten, as well as Pfungst himself and other informed questioners, all involuntarily bent their head forward when Hans was to start, and involuntarily bent their head slightly backwards when Hans had reached the correct number. With his investigation, **Pfungst identified a potentially major disruptive factor of scientific work, a special type of observer bias which became known as the Clever Hans Effect.** This is a serious and almost unbelievable old example, but is very much relevant in contemporary animal behaviour research (Lit et al. 2011; Chu et al. 2022), for instance in the area of animal cognition (Umiker-Sebeok and Sebeok 1981; Samhita and Gross 2013). Fortunately, aware researchers now commonly take precautions against this phenomenon, for example by only using observers blind to the experimental condition (Horowitz et al. 2013; Proops et al. 2018), or using innovative methods that circumvent human handlers altogether (Krause 2016; Dudde et al. 2018; Hopper et al. 2019; Gatto et al. 2020).

References

Brumm H, Naguib M (2009) Environmental acoustics and the evolution of bird song. Adv Study Behav 40:1–33

Brumm H, Zollinger SA, Niemela PT, Sprau P (2017) Measurement artefacts lead to false positives in the study of birdsong in noise. Methods Ecol Evol 8:1617–1625

Caro TM, Roper R, Young M, Dank GR (1979) Inter-observer reliability. Behaviour 69:303–315

Chu PC, Wierucka K, Murphy D, Tilley HB, Mumby HS (2022) Human interventions in a behavioural experiment for Asian Elephants (*Elephas maximus*). Anim Cogn 26:393–404

Cohen J (1968) Weighted kappa – nominal scale agreement with provision for scaled disagreement or partial credit. Psychol Bull 70:213–220

Dudde A, Krause ET, Matthews LR, Schrader L (2018) More than eggs–relationship between productivity and learning in laying hens. Front Psychol 9:2000

Ferald LD (1983) The Hans Legacy: a story of science. Erlbaum, New Jersey

Gatto E, Lucon-Xiccato T, Bisazza A, Manabe K, Dadda M (2020) The devil is in the detail: Zebrafish learn to discriminate visual stimuli only if salient. Behav Proc 179:104215

Hopper LM, Egelkamp CL, Fidino M, Ross SR (2019) An assessment of touchscreens for testing primate food preferences and valuations. Behav Res Methods 51:639–650

Horowitz A, Hecht J, Dedrick A (2013) Smelling more or less: investigating the olfactory experience of the domestic dog. Learn Motiv 44:207–217

Jones AE, Ten Cate C, Bijleveld CJH (2001) The interobserver reliability of scoring sonagrams by eye: a study on methods, illustrated on zebra finch songs. Anim Behav 62:791–801

Krause ET (2016) Colour cues that are not directly attached to the body of males do not influence the mate choice of Zebra Finches. PLoS ONE 11:e0167674

Lit L, Schweitzer JB, Oberbauer AM (2011) Handler beliefs affect scent detection dog outcomes. Anim Cogn 14:387–394

Lloyd P, Plagányi ÉE (2002) Correcting observer effect bias in estimates of nesting success of a coastal bird, the White-fronted Plover *Charadrius marginatus*. Bird Study 49:124–130

Mayfield H (1961) Nesting success calculated from exposure. The Wilson Bulletin 73:255–261

McGrath JC, Drummond GB, McLachlan EM, Kilkenny C, Wainwright CL (2010) Guidelines for reporting experiments involving animals: the ARRIVE guidelines. Br J Pharmacol 160:1573–1576

Nickerson RS (1998) Confirmation bias: A ubiquitous phenomenon in many guises. Rev Gen Psychol 2:175–220

Pfungst O (1907) Das Pferd des Herrn von Osten: dDer kluge Hans Ein Beitrag zur experimentellen Tier-und Menschen-Psychologie. Barth, Leipzig

Proops L, Grounds K, Smith AV, McComb K (2018) Animals remember previous facial expressions that specific humans have exhibited. Curr Biol 28:1428–1432

Rodway MS, Montevecchi WA, Chardine JW (1996) Effects of investigator disturbance on breeding success of Atlantic puffins *Fratercula arctica*. Biol Cons 76:311–319

Rosenthal R, Fode KL (1963) The effect of experimenter bias on the performance of the albino rat. Behav Sci 8:183–189

Samhita L, Gross HJ (2013) The "clever Hans phenomenon" revisited. Communicative & Integr Biol 6:e27122

Slabbekoorn H, Peet M (2003) Birds sing at a higher pitch in urban noise. Nature 424:269

Strang CA (1980) Incidence of avian predators near people searching for waterfowl nests. J Wildl Manag 44:220–222

Tuyttens FAM, de Graaf S, Heerkens JL, Jacobs L, Nalon E, Ott S, Stadig L, van Laer E, Ampe B (2014) Observer bias in animal behaviour research: can we believe what we score, if we score what we believe? Anim Behav 90:273–280

Umiker-Sebeok J, Sebeok TA (1981) Clever Hans and smart simians: The self-fulfilling prophecy and kindred methodological pitfalls. Anthropos 76:89–165

Webster MM, Rutz C (2020) How STRANGE are your study animals? Nature 582:337–340

Whitney WO, Mehlhaff CJ (1987) High-rise syndrome in cats. J Am Vet Med Assoc 191:1399–1403

5

Tools for Measuring Behaviour

Contents

© The Author(s) and Friedrich-Loeffler-Institut, under exclusive license to Springer-Verlag
GmbH, DE, part of Springer Nature 2023
M. Naguib et al., *Methods in Animal Behaviour*,
https://doi.org/10.1007/978-3-662-67792-6_6

6

The use of tools is ubiquitous in modern animal behaviour studies. At minimum, behavioural observations are regularly recorded using a computer or tablet, and frequently also with **audio and video recording devices.** Other electronic tools are a prerequisite for recording and accessing certain behaviours for analysis. Such tools have the advantage that they can record more data, often in greater detail, than human observers, as well as in situations that are not possible for humans to observe. In addition, automated data collection and behavioural analysis tools are being used more and more frequently (Vogt 2021). In addition to saving time, automatic procedures have the advantage that the registration of data is no longer subject to immediate human error and reduce the potential for subjective or observer-specific effects (e.g. observer bias, ▶ Chap. 5). Furthermore, automated data collection enables a replicable standardisation of data registration. Validation of the automated methods is an important prerequisite for this, because even automated registration systems have certain measurement errors (Royer and Lutcavage 2008; Russo and Voigt 2016; Wurtz et al. 2019) and the analysis of such data requires a range of additional, often subjective decisions, such as when thresholds and filters are being applied to the raw data.

Data loggers with specific sensors attached to the animal make it possible to automatically record and store relevant information at specific time intervals, such as activity, acceleration values, and location (Jetz et al. 2022; Smith and Pinter-Wollman 2021). There are many different sensor types, not only for animal behaviour (e.g. movement) but also for animal physiology (e.g. heartbeat) and local environment (e.g. temperature), which can inform animal behaviour and even may be integrated into a single device. Data from multiple different sensors can be combined, synchronised, and processed using low-cost **single-board computers** (e.g. Raspberry Pi) or **microcontrollers (e.g.** Arduino (Jolles 2021)).

Transmitter-receiver systems (Hughey et al. 2018; Smith and Pinter-Wollman 2021; Griebling et al. 2022) such as radio-frequency identification **(RFID) transponders (Fig.** 6.1 **and Sect** 2.6.1**)** or **satellite** (e.g. global positioning system, GPS) **transmitters (Fig.** 6.2**)** allow identification or localisation, respectively. These systems can be **passive** or **active**. Passive transponders (e.g. RFID tags) get power from a received signal and only transmit a signal in response to a received signal. Active transmitters (e.g. GPS tags) have their own power supply and can transmit signals independently. However, every automated data acquisition system has a certain measurement error; for example, how well RFID antennas (◘ Fig. 6.3) detect the passive transponders may depend on the angle of the transponder relative to the receiver antenna and on the environmental conditions. If RFID tags are used to determine a feeder and a nesting site, and a detection is being missed, the data can be substantially 'off' as the subsequent presence—absence detections will be off. For example, an animal may being recorded as still present on a site even though it has left and then recorded as leaving when it passes through the antenna on the next visit. A researcher should thus always determine the accuracy and validity of electronic tools before employing them in a study.

◨ **Fig. 6.1**   Example of a small passive RFID transponder

◨ **Fig. 6.2**   Condor with wing tags and transmitter

In the following section, we highlight a few digital tools, i.e. video, audio and GPS/radio tags, through examples, but this overview is by no means comprehensive. Given the exponential growth in the variety and capabilities of user-friendly technology, we advise to get up-to-date with the most recent advancements before starting your study.

6

6.1 Video Recording, Video Playback and Photos

The use of video recordings is particularly helpful to study behaviours of animals that are easily disturbed by human presence or live in habitats that are not accessible or impractical for direct observation. Moreover, video recordings allow for more detailed analyses and observer reliability testing (► Chap. 5) at a later stage. Videos may be set to record at a defined time or during fixed intervals. However, it is also possible to automatically record certain events (e.g. via motion detection), which is helpful, for example, in determining activity rhythms of nocturnal animals. Videos can also allow for non-invasive video-tracking of multiple animals simultaneously. Such approaches are used to study the collective behaviour of animals in great detail, such as the anti-predator responses of fish in schools (Hughey et al. 2018; Smith and Pinter-Wollman 2021). When a reliable tracking algorithm is implemented, live observers no longer have to make ad hoc decisions on whether a behaviour fits a specific category. Complete behavioural sequences can be automatically categorized, archived, and carefully checked without immediate time-pressure. Furthermore, video recordings make it possible to use 'blind' observers, i.e. people who are unaware of the experimental conditions, or multiple observers, and so minimize observer bias in your study (► Chap. 5).

Analyses of video sequences can, however, be very time-consuming when they are not automated, so it should be well thought through when video technology is preferred over direct observation. This may be the case when behaviours are too frequent to reliably score in real-time. Moreover, with video recording, we run the risk of accumulating massive volumes of data without having made a clear decision on which data should be analysed and how. However, observers can save an enormous amount of time by fast-forwarding through sequences of little interest.

Video recordings may not be sufficient as the sole technique, for example, when studying social interactions of group-living animals in the wild. If the entire behavioural context cannot be captured, it can lead to a lack of important contextual information, like other group members, missing from the analysis. Yet, using a too wide view can lead to missing important details about the individual behaviour of a focal animal. It is thus crucial to have the research question clear as well as they type of data required to answer it prior to applying and programming the cameras. Special high-speed cameras can be useful for recording very fast behaviour such as flight movements, leg coordination, or sound production, and these behaviours can then made accessible for analysis via slow motion or single-frame analysis (Podos et al. 2004; Eckmeier et al. 2008; Tomotani and Muijres 2019). Video technology holds enormous potential for the study of animal behaviour, including on time and spatial scales that are normally out of reach for human observers.

Videos can also be used as experimental stimuli (Oliveira et al. 2000; Choudinard-Thuly et al. 2017; Snijders et al. 2017a). Video playbacks, computer animations, and virtual reality offer the possibility to precisely define and replicate an experimental stimulus. To maximize the realism of the perceived images, there are several important considerations, including minimizing light reflection, maximizing resolution, and taking into account an animal's flicker-fusion frequency (i.e. the frequency at which a flashing light is perceived as constant). In addition, many animals perceive colours differently than humans or have extended colour perception (e.g. the UV range). This means video playback may not be suitable for every study organism or the video may need to be altered to match the viewer's (i.e. the study animal's) colour perception (Woo and Rieucau 2011).

In addition to video, photos are often used to monitor animals (e.g. camera traps; Burton et al. 2015; Caravaggi et al. 2020). Due to their static nature it can be more challenging to use photos to study dynamic behaviour, but they can be useful to identify individual animals, infer spatial distribution, or get morphological information.

Digital cameras are primarily adapted to the human perception spectrum, which can lead us to miss information that is important to the animals. Therefore, there are special cameras that focus on information that may be more important to the sensory ecology of your study animal, such as UV, polarized light, or infra-red. Such specialized cameras can also be used for different types of research questions, such as thermal imaging cameras that can inform researchers on animal physiology and health (◘ Fig. 6.4).

6.2 Recording, Analysis and Playback of Acoustic Data (Bioacoustics)

Many animal species use acoustic signals for communication. Such signals often play a role in the recruitment and defence of resources such as territories and mating partners, predator avoidance, foraging, and parent-offspring communication (e.g. Janik 2009; Zuberbühler 2009; Bradbury and Vehrencamp 2011; Naguib

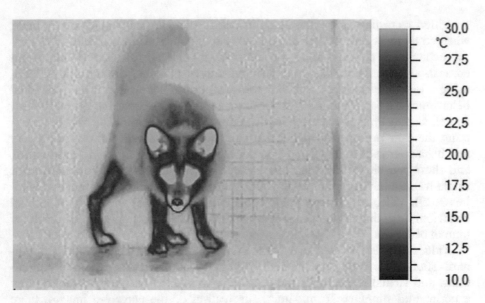

Fig. 6.4 Heat camera image of a red fox (*Vulpes vulpes*). The colour scale shows the heat radiation emitted by the animal. One can easily see the insulating effect of the fur on the body. (Photo: Verena Schöler, E. Tobias Krause, Conrad M. Freuling, Thomas Müller, Friedrich-Loeffler-Institut)

and Riebel 2014), to name but a few contexts. Moreover, acoustic variation can provide eavesdropping conspecifics (or researchers) information about the age, size, motivation, personality, stress, or population membership of the sender (McGregor 2005; Snijders and Naguib 2017).

The **acoustic signals produced by organisms can usually be accurately recorded** with the help of audio devices. The methods for recording and analysing bioacoustics are too comprehensive to discuss in detail but we will provide a short overview. In brief, **animal communication researchers use both handheld recorders to target specific individuals, or automatic stationary recorders when locations of vocalization can be predicted (such as near nests). Automatic stationary recorder are especially helpful to collect standardized detections of animal presence across seasons or to monitor the soundscape of a habitat** (Loning et al. 2023). Animal-borne recorders are also available (Gill et al. 2015; Cvikel et al. 2015), which are attached to the animals themselves. Moreover, entire **microphone arrays** can be used to track the location of sound-producing animals (Foote et al. 2010; Spillmann et al. 2015). Overall, recording acoustic signals with sensitive audio equipment rather than live scoring allows us to study animal communication, specifically the presence of vocalizing animals and the temporal structure and phonological characteristics, in much greater detail.

When selecting the recording devices it is important to ensure that the input level can be manually adjusted and is turned on. Automatic regulation of the input level, as is done by many simple devices, can lead to a change in the audio volume structure of the sounds to be recorded. Background **noise** is another common challenge with audio recordings, especially in the field (Brumm 2013; Naguib 2013). In order to estimate the quality of recordings, we advise you to inspect the sound spectrograms of your first recordings. This allows you to estimate whether or not the quality is sufficient for the desired purpose before you continue with the recordings. Background noise can interfere with the reliability of certain acoustic parameters during analysis or can partially mask signals during playback experiments. Moreover, when recordings are made from a long distance, degradation of sound that accumulates over distance will lead to changes in the signal, such as attenuated high frequencies, modified amplitude structure and reverberation. Such signal degradation can impair the analysis of the sounds and using degraded signals as stimuli in playback experiments can have strong effects on how animals respond (Wiley and Richards 1978; Naguib and Wiley 2001).

A detailed analysis of sounds allows us to quantify details in signals, like high frequency content or rapid changes in frequency or amplitude, and often such details carry important information. A number of computer programs are available for the analysis of acoustic signals, including free and open source options. This makes acoustic analysis comparatively easy to access. However, we must be careful not to blindly rely on pre-defined routines or settings in computer programs, since the settings for signal digitisation and sound spectrogram calculation affect the data structure and default settings often are tuned towards human perception. Therefore, before starting an analysis, one should become familiar with the basic principles of acoustics, including the sampling rates and the effect Fast Fourier Transformation (FFT) settings for spectrogram time and frequency resolution. This will allow you to adjust the settings for spectrogram calculation and sound analysis in a way that you expect to be meaningful for your question and the study species. **The most important parameters of acoustic signals are the amplitude (volume; measured in decibels), the frequency (pitch; number of oscillations per second; measured in Hertz) and the temporal structure.**

Visual representations can be important for measuring the temporal parameters of a sound. Common visual representations of acoustic signals include the spectrogram and oscillogram (◘ Fig. 6.5). An **oscillogram**, or envelope curve, is comprised of the amplitude plotted over time, but gives no information on the frequency structure of the sound. The **Fast Fourier Transformation** (FFT) can be used to deconstruct individual frequencies in a certain time window (as it is done by our ear). Many successive FFTs then lead to a **sound spectrogram** (◘ Fig. 6.5) in which the frequency structure over time is represented and this representation visualizes a sound corresponding to how we hear it.

When interpreting differences revealed by an analysis of acoustic signals, it is ultimately useful to consider the characteristics of sound production as well as the perceptual and processing abilities of the species concerned. We cannot conclude from pure analysis whether differences are meaningful if we do not also test the response of receivers to the sounds. Even where differences are not detectable for us, categorical perception can allow animals to distinguish sounds on the basis of characteristics that do not emerge as distinct (Caves et al. 2019; Peniston et al. 2020). A good visual example of this is the rainbow, which has a continuous

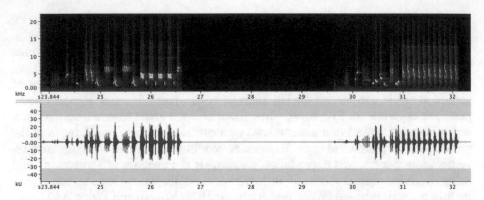

6

☐ **Fig. 6.5** Visualization of two songs of a common nightingale (*Luscinia megarhynchos*), shown above as a sound spectrogram (sonagram) and below as an oscillogram. A sound spectrogram shows the frequency (in kHz) on the y-axis against time with the volume (amplitude) as colour coding, while the oscillogram shows the volume against time. Oscillograms are particularly well suited for measuring the time structure of a sound, while spectrograms are used for structural analysis

light spectrum but is perceived by us as consisting of discretely different colours. Through sound analysis, often in combination with additional data on the environment, context, and individual, you can thus learn more about the structures and function of signals, the context in which these sounds are used (e.g. predator detection), the different factors that drive variation in acoustic traits, or if and how acoustic communication is learned.

Acoustic signals can be passively recorded but can also be broadcast to study animal behaviour. In many taxonomic animal groups, the playback of sounds (**playback experiments**) is used as a key method to gain insights into the mechanisms or functions of acoustically-mediated behaviours (McGregor 2000; King 2015), such as territory defence or group coordination. The use of audio equipment with often many options to adjust the settings requires some basic knowledge of bioacoustics and the species (e.g. what is the relevant frequency range). Animals sometimes have different perception thresholds compared to us, so that the recording, analysis, and playback of acoustic signals in playback experiments should be carried out according to some basic standards. For example, it is best to avoid audio compression procedures that are specifically adapted to human hearing, such as mp3 conversions in bioacoustics resarch.

6.3 Animal Movement Tracking

Technological advances in recent years have created an abundance of new tools to track the movement of animals. While there are many 'biologging' devices that can record information about an animal's behaviour, physiology, cognition, and abiotic conditions, this section focuses on devices that are used to determine an animal's position in space and time. There are now many tools to track the movement of individual animals. These tracking devices enable behavioural biologists to address new questions that aim to understand the importance of how and why an individual animal moves (Nathan et al. 2008). Many of these tools are designed to track animals in the wild (e.g. geolocator, radio-tags or GPS tags), but some may also be used to track animals in captivity on smaller scales (e.g., passive integrated RFID transponder coupled with RFID readers). Below are some of the major features associated with each of these tracking devices that are important to consider when designing an animal tracking study.

Tracking devices may be broadly classed as loggers or transmitters (also called biotelemetry). Biologging devices store data onboard and thus must later be recovered to download the data. Some examples of biologgers are geolocators, many GPS tags, time-depth recorders (TDRs), and inertial sensors (Lahoz-Monfort and Magrath 2021). Since these tags must be retrieved to recover the data (unless coupled with a remote download option), these devices are only suitable for animals that can be reliably recaptured. Numerous factors will affect whether a tagged individual is likely to be recaptured, such as dispersal tendency, territoriality, whether individuals become 'trap-shy', and mortality, which can always occur, but may be more likely to affect certain life-stages. A primary advantage of biologgers, however, is that since the data is not transmitted, the energy consumption of these tags is low compared to transmitters, and so they can be deployed on a wider range of species (Nathan et al. 2022).

Transmitters, unlike loggers, do not necessarily need to be recovered from the tagged animal to retrieve the data since they are broadcast via e.g. satellite, cellular, or a local wireless network (Lahoz-Monfort and Magrath 2021). Examples of transmitter tags are Platform Transmitting Terminal (PTT), some GPS, and radio tags. Tags that transmit via satellite systems (such as Argos or ICARUS (Wikelski et al. 2007)) are able to relay data from virtually anywhere on the planet and so are suitable for tracking species in remote regions. Tracking devices may also transmit via the Global System for Mobile Communications (GSM) network, which allows data to be remotely accessed, but only where there is coverage. Local wireless networks can also be used to download data from tracking tags though these systems require that the tagged individual is within a known area (e.g. a breeding colony). For example, some GPS tags can upload data to a stationary base station (e.g. UvA-BiTS) or via a mobile remote downloading device, such as a drone (Jin et al. 2023). Radio tags are also transmitters, but unlike most other tracking devices (e.g. satellite, GPS, geolocators), the localisation of the tagged animal generally requires an additional step by the researcher to process the radio signals from the tag, typically using either triangulation or trilateration. Most radio tracking studies have used handheld directional antenna, however, automated radio tracking with fixed receivers is increasingly being applied (Kays et al. 2011; Snijders et al. 2017b). Modern radio tags can be used to track small species (e.g. insects: Fisher et al. 2021), however, additional infrastructure is needed in the form of receivers to detect signals.

Box 6.1: Automated and Artificial Intelligence (AI) Based Methods to Register Animal Behaviour

Automatic registration methods are particularly suitable for activity measurements and the registration of spatial behaviour (Kjaer 2017; ◘ Fig. 6.3). Automated recognition of detailed behaviour is often still difficult, but becomes more feasible with the rapid developments in machine learning and neural networks (Valletta et al. 2017; Gutierrez-Galan et al. 2018; Wang 2019; Schütz et al. 2022; Marks et al. 2022). In principle, many methods for automated registration of behaviour are conceivable. They can minimise problems such as observer bias, but also need to be sufficiently validated before they can be used.

For example, single frames from video recordings (Schütz et al. 2021) can be analysed using YOLOv4 (You only look once). YOLOv4 is a computer vision architecture that can make object detection in real-time using convolutional neural networks (CNN) (Schütz et al. 2021) (◘ Fig. 6.6).

(a) Night scene.   (b) Day scene.

(c) Night scene.   (d) Day scene.

◘ **Fig. 6.6**   Silverfoxes are automatically detected using YOLOv4 algorithms. (From Schütz et al. 2021; CC-BY 4.0 license, ▶ https://doi.org/10.3390/ani11061723)

Given the increasing miniaturization of such tracking tools, an ever-growing number of species are suitable for tracking (Nathan et al. 2022). Fundamental constraints on sampling regimes for a given tag size, however, necessitates that researchers carefully consider their research questions before choosing a tracking device. In essence, greater power consumption requires larger batteries (or solar power, which also increases tag weight) and so greater operation time (either longevity or frequency) will typically require heavier tags. Correspondingly, device capabilities (e.g. satellite or GSM connectivity) also typically scale with size. For example, GPS tags with satellite or GSM connectivity are capable of tracking animals across the globe with high spatial and temporal resolution for extended periods of time, yet currently they can only be used on 19% and 14% of bird species, respectively (Nathan et al. 2022). In contrast, radio tags can be deployed on over 90% of bird species, yet these tags require additional receivers and so only work within a relatively limited spatial extent and have comparably low spatial resolution. Cost is another consideration that scales with device functionality and can limit the number of individuals that can be concurrently tracked. Applying any device to an animal maybe be considered as animal experiment in some countries, so will require ethical permission from the relevant authorities. Next to this formal procedure, it is always important to assess to what extent a device will affect the behaviour (or even survival) of the tagged animal (Barron et al. 2010; Snijders et al 2017c).

References

Barron DG, Brawn JD, Weatherhead PJ (2010) Meta-analysis of transmitter effects on avian behaviour and ecology. Methods Ecol Evol 1:180–187

Bradbury JW, Vehrencamp SL (2011) Principles of animal communication. Sinauer Associates, Sunderland

Brumm H (2013) Animal communication and noise, vol 2. Springer Science & Business Media, Berlin

Burton AC, Neilson E, Moreira D, Ladle A, Steenweg R, Fisher JT, Bayne E, Boutin S (2015) Wildlife camera trapping: a review and recommendations for linking surveys to ecological processes. J Appl Ecol 52:675–685

Caravaggi A, Burton C, Clark DA, Fisher JT, Grass A, Greem S, Hobaiter C, Hofmeester TR, Kalan AK, Rabaiotti D, Rivet D (2020) A review of factors to consider when using camera traps to study animal behavior to inform wildlife ecology and conservation. Conservat Sci Prac 2:e239

Caves EM, Nowicki S, Johnsen S (2019) Von Uexkull revisited: addressing human biases in the study of animal perception. Integr Comp Biol 59:1451–1462

Chouinard-Thuly L, Gierszewski S, Rosenthal GG, Reader SM, Rieucau G, Woo KL, Gerlai R, Tedore C, Ingley SJ, Stowers JR, Frommen JG, Dolins FL, Witte K (2017) Technical and conceptual considerations for using animated stimuli in studies of animal behavior. Curr Zool 63:5–19

Cvikel N, Berg KE, Levin E, Hurme E, Borissov I, Boonman A, Amichai E, Yovel Y (2015) Bats aggregate to improve prey search but might be impaired when their density becomes too high. Curr Biol 25:206–211

Eckmeier D, Geurten BRH, Kress D, Mertes M, Kern R, Egelhaaf M et al (2008) Gaze strategy in the free flying zebra finch (Taeniopygia guttata). PLoS ONE 3:e3956

Fisher KE, Dixon PM, Han G, Adelman JS, Bradbury SP (2021) Locating large insects using automated VHF radio telemetry with a multi-antennae array. Methods Ecol Evol 12:494–506

Foote JR, Fitzsimmons LP, Mennill DJ, Ratcliffe LM (2010) Black-capped chickadee dawn choruses are interactive communication networks. Behaviour 147:1219–1248

Gill LF, Goymann W, Ter Maat A, Gahr M (2015) Patterns of call communication between group-housed zebra finches change during the breeding cycle. elife 4:e07770

Griebling HJ, Sluka CM, Stanton LA, Barrett LP, Bastos JB, Benson-Amram S (2022) How technology can advance the study of animal cognition in the wild. Curr Opin Behav Sci 45:101120

Gutierrez-Galan D, Dominguez-Morales JP, Cerezuela-Escudero E, Rios-Navarro A, Tapiador-Morales R, Rivas-Perez M, Dominguez-Morales M, Jimenez-Fernandez A, Linares-Barranco A (2018) Embedded neural network for real-time animal behavior classification. Neurocomputing 272:17–26

Hughey LF, Hein AM, Strandburg-Peshkin A, Jensen FH (2018) Challenges and solutions for studying collective animal behaviour in the wild. Philos Trans R Soc B: Biol Sci 373:20170005

Janik VM (2009) Acoustic communication in delphinids. Adv Study Behav 40:123–157

Jetz W, Tertitski G, Kays R, Mueller U, Wikelski M (2022) Biological Earth observation with animal sensors. Trends Ecol Evol 37:293–298

Jin T, Si X, Liu J, Ding P (2023) An integrated animal tracking technology combining a GPS tracking system with a UAV. Methods Ecol Evol 14:505–511

Jolles JW (2021) Broad-scale applications of the Raspberry Pi: a review and guide for biologists. Methods Ecol Evol 12:1562–1579

Kays R, Tilak S, Crofoot M, Fountain T, Obando D, Ortega A, Kuemmeth F, Mandel J, Swenson G, Lambert T, Hirsch B, Wikelski M (2011) Tracking animal location and activity with an automated radio telemetry system in a tropical rainforest. Comput J 54:1931–1948

King SL (2015) You talkin'to me? Interactive playback is a powerful yet underused tool in animal communication research. Biol Let 11:20150403

Kjaer JB (2017) Divergent selection on home pen locomotor activity in a chicken model: selection program, genetic parameters and direct response on activity and body weight. PLoS ONE 12:e0182103

Lahoz-Monfort JJ, Magrath MJ (2021) A comprehensive overview of technologies for species and habitat monitoring and conservation. Bioscience 71:1038–1062

Loning H, Verkade L, Griffith SC, Naguib M (2023) The social role of song in wild zebra finches. Curr Biol 33:372–380

Marks M, Jin Q, Sturman O, von Ziegler L, Kollmorgen S, von der Behrens W, Mante V, Bohacek J, Yanik MF (2022) Deep-learning-based identification, tracking, pose estimation and behaviour classification of interacting primates and mice in complex environments. Nat Mach Intell 4:331–340

McGregor PK (2000) Playback experiments: design and analysis. Acta Etologia 3:3–8

McGregor PK (2005) Communication networks. Cambridge University Press, Cambridge

Nathan R, Getz WM, Revilla E, Holyoak M, Kadmon R, Saltz D, Smouse PE (2008) A movement ecology paradigm for unifying organismal movement research. Proc Nat Acad Sci 105:19052–19059

Nathan R, Monk CT, Arlinghaus R, Adam T, Alós J, Assaf M, et al (2022) Big-data approaches lead to an increased understanding of the ecology of animal movement. Sci 375:eabg1780.

Naguib M, Wiley RH (2001) Estimating the distance to a source of sound: mechanisms and adaptations for long-range communication. Anim Behav 62:825–837

Naguib M (2013) Living in a noisy world: indirect effects of noise on animal communication. Behaviour 150:1069–1084

Naguib M, Riebel K (2014) Singing in space and time: the biology of birdsong. In: Witzany G (ed) Biocommunication of animals. Springer, Dordrecht, pp 233–247

Oliveira RF, Rosenthal GG, Schlupp I, McGregor PK, Cuthill IC, Endler JA, Fleishman LJ, Zeil J, Barata E, Burford F, Goncalves D, Haley M, Jakobsson S, Jennions MD, Körner KE, Lindsström L, Peake T, Pilastro A, Pope DS, Roberts SGB, Rowe C, Smith J, Waas JR (2000) Considerations on the use of video playbacks as visual stimuli: the Lisbon workshop consensus. Acta Ethologica 3:61–65

Peniston JH, Green PA, Zipple MN, Nowicki S (2020) Threshhold assessment, categorical perception, and the evolution of reliable signaling. Evolution 74:2591–2604

Podos J, Southall JA, Rossi-Santos MR (2004) Vocal mechanics in Darwin's finches: correlation of beak gape and song frequency. J Exp Biol 207:607–619

Royer F, Lutcavage M (2008) Filtering and interpreting location errors in satellite telemetry of marine animals. J Exp Mar Biol Ecol 359:1–10

Russo D, Voigt CC (2016) The use of automated identification of bat echolocation calls in acoustic monitoring: a cautionary note for a sound analysis. Ecol Ind 66:598–602

Schütz AK, Schöler V, Krause ET, Fischer M, Müller T, Freuling CM, Conraths FJ, Stanke M, Homeier-Bachmann T, Lentz HHK (2021) Application of YOLOv4 for detection and motion monitoring of red foxes. Animals 11:1723

Schütz AK, Krause ET, Fischer M, Müller T, Freuling CM, Conraths FJ, Homeier-Bachmann T, Lentz HHK (2022) Computer vision for detection of body posture and behavior of red foxes. Animals 12:233

Smith JE, Pinter-Wollman N (2021) Observing the unwatchable: integrating automated sensing, naturalistic observations and animal social network analysis in the age of big data. J Anim Ecol 90:62–75

Snijders L, Naguib M (2017) Communication in animal social networks: a missing link? Adv Study Behav 49:297–359

Snijders L, Naguib M, van Oers K (2017a) Dominance rank and boldness predict social attraction in great tits. Behav Ecology 28:398–406

Snijders L, van Oers K, Naguib M (2017b) Sex-specific responses to territorial intrusions in a communication network: evidence from radio-tagged great tits. Ecol Evol 7:918–927

Snijders, L., Weme, L. E. N., de Goede, P., Savage, J. L., van Oers, K., & Naguib, M. (2017c). Context-dependent effects ofradio transmitter attachment on a small passerine. Journal of Avian Biology, 48(5), 650–659. ► https://doi.org/10.1111/jav.01148

Spillmann B, van Noordwijk MA, Willems EP, Mitra Setia T, Wipfli U, van Schaik CP (2015) Validation of an acoustic location system to monitor Bornean orangutan (*Pongo pygmaeus wurmbii*) long calls. Am J Primatol 77:767–776

Tomotani BM, Muijres FT (2019) A songbird compensates for wing molt during escape flights by reducing the molt gap and increasing angle of attack. J Exp Biol 222:jeb195396

Valletta, JJ, Torney, C, Kings, M, Thornton, A, & Madden, J (2017) Applications of machine learning in animal behaviour studies. Animal Behaviour 124:203–220.

Vogt N (2021) Automated behavioral analysis. Nat Methods 18:29

Wang G (2019) Machine learning for inferring animal behavior from location and movement data. Eco Inform 49:69–76

Wikelski M, Kays RW, Kasdin NJ, Thorup K, Smith JA, Swenson GW (2007) Going wild: what a global small-animal tracking system could do for experimental biologists. J Exp Biol 210:181–186

Wiley RH, Richards DG (1978) Physical constraints on acoustic communication in the atmosphere: implications for the evolution of animal vocalizations. Behav Ecol Sociobiol 3:69–94

Woo KL, Rieucau G (2011) From dummies to animations: a review of computer-animated stimuli used in animal behavior studies. Behav Ecol Sociobiol 65:1671–1685

Wurtz K, Camerlink I, D'Eath RB, Fernández AP, Norton T, Steibel J, Siegford J (2019) Recording behaviour of indoor-housed farm animals automatically using machine vision technology: a systematic review. PLoS ONE 14:e0226669

Zuberbühler K (2009) Survivor signals: the biology and psychology of animal alarm calling. Adv Study Behav 40:277–322

Data Analysis and Presentation

Contents

© The Author(s) and Friedrich-Loeffler-Institut, under exclusive license to Springer-Verlag
GmbH, DE, part of Springer Nature 2023
M. Naguib et al., *Methods in Animal Behaviour*,
https://doi.org/10.1007/978-3-662-67792-6_7

7.1   Statistical Data Analysis

In animal behaviour research, a comprehensive understanding of how to handle your data and their statistical analysis is essential. Statistical tests allow you to draw scientific conclusions from your data. In this chapter we will discuss only some of the very basic aspects of data analysis. For detailed guidance of the most appropriate procedures as well as for more advanced statistical tests, we advise to consult dedicated statistical textbooks (e.g. Quinn and Keough 2002; Zar 2013; Jones et al. 2022).

Basically, we can make a distinction between **descriptive and inferential statistics**. With descriptive statistics we calculate and display the features of a dataset. Descriptive statistics describe the data, but you cannot (yet) draw firm conclusions from them. To draw conclusions about whether the data reveal a biologically meaningful difference or relationship, rather than random variation, we need to use inferential statistics (▶ Sect. 7.1.2). Nevertheless, descriptive statistics are important and informative. For example, knowing the distribution of the data will help to select the most appropriate statistical test. Almost all statistical tests have a number of assumptions that the data will need to fulfil or otherwise the test outcome may not be reliable. Descriptive statistics help to evaluate these assumptions. Moreover, while inferential statistics inform about whether a difference or correlation is significant, descriptive statistics let us know about the size of the difference or strength of the correlation (i.e. the effect size). Data analyses thus should include both descriptive and inferential statistics.

7.1.1   Descriptive Statistics

Descriptive statistics include measures of central tendency (also known as averages, e.g. mean, median, mode, ▶ Box 7.1), which can later tell us something about the effect size, or strength, of the relationships between the variables. However, a single measure of the central value is insufficient to represent the nature of the data. Therefore, measures of centrality should always be combined with at least one measure of dispersion. Such measures indicate how spread out the individual observations in the dataset are. There are several measures of data dispersion to choose from (e.g. range, variance, standard deviation, standard error, 95% confidence interval, ▶ Box 7.2). The decision about which of these measures to choose depends on several aspects including the type of data (e.g. nominal, ordinal, interval, or ratio) and the distribution of the data, which could be, for example, skewed or normally distributed, i.e. bell-shaped (◘ Fig. 7.1).

Box 7.1: Different Measures of Central Tendency (Average Values)

The most common measures of central values of data are:

- **Arithmetic Mean**: Usually simply called the mean value. The mean is the sum of the data values divided by the number of observations. This is a good measure of central tendency when the data are normally distributed.
- **Median**: The median represents the value that is in the middle when all of the observation values are ordered from smallest to largest. This is an appropriate measure when the data are not normally distributed and is also useful for ordinal (ranked) data.
- **Mode**: The mode is the most frequent value. This is the only measure of central tendency that can be used for nominal data (qualitative categories) and is also useful for ordinal data.

Example: You collected a dataset of seven observations on the number of birds visiting feeders: 5, 2, 100, 4, 2, 5, 2. In this case, the measures of the central position will be:

- **Arithmetic Mean**: $(5 + 2 + 100 + 4 + 2 + 5 + 2) / 7 = 17.14$
- **Median**: 2, 2, 2, **4**, 5, 5, $100 = 4$
- **Mode**: $3 \times 2, 1 \times 4, 2 \times 5, 1 \times 100 = 2$

The data in this example corresponds to a right-skewed distribution (◘ Fig. 7.1) with many small values (e.g. 2, 4, 5) and a few very large values (e.g. 100). **In such a skewed distribution, the three measures of central tendency substantially differ from one another**. Here, the arithmetic mean is not a good measure for summarising the data because it is strongly influenced by the one extreme value of 100. The median would therefore be a more suitable measure.

Box 7.2: Dispersion Measures

There are several different measures of the dispersion, or the extent of variability, of the data. The decision of which measures to calculate depends mainly on the distribution of the data.

- **Range**: This is the difference between the smallest and largest value in the data. So, for the observation values in ▶ Box 7.1 the range would be 98 (100-2). Be careful, because this measure is strongly influenced by outliers and thus may not be a very meaningful representation of your data.
- **Variance**: The variance is calculated by taking the average of the squared differences from the Mean. In steps: first, subtract each value from the mean value and square each result. Then, add all the resulting values together and divide by the number of observations (sample size). Now you have the variance. Note that since the Variance is calculated from the Mean, it is only suitable for data distributions for which the Arithmetic Mean is indeed an appropriate measure.

— **Standard Deviation**: This is a commonly used measure in biology when describing the variation of actually collected data. It gives an indication of how much the observation values deviate from the mean value. The Standard Deviation (S.D.) is calculated as the square root of the Variance. As with Variance, this measure is only suitable for data distributions for which the Arithmetic Mean is appropriate, such as for symmetrical distributions, like the normal (bell-shaped) distribution.

— **Standard Error**: This measure estimates how close the Mean of your dataset is to the true Mean of the population. The population mean is the average value of the entire population. Because we often cannot measure the entire population, we work with a sample Mean. One calculates the Standard Error (S.E.) by taking the Standard Deviation and divide it by the square root of the number of observations (sample size). Again, the Standard Error is most appropriate for relatively symmetrical distributions and usually used along with inferential statistics.

— **95% Confidence Interval**: The 95% Confidence Interval estimates the range of values within which the true population Mean would lie. However, there is still a 5% chance that the true population Mean does not lie in this range (hence, 95%). It is thus a measure of certainty, or margin of error, around the Mean. To calculate the 95% Confidence Interval you need the so-called Z value (1,960 for 95%) and multiply this by the Standard Error. For the lower confidence value, one subtracts the result from the Mean and for the upper value, you add the result to the Mean. Again, this measure is most appropriate for relatively symmetrical distributions. There are ways to calculate the 95% Confidence Interval for non-normally distributed data, but this requires more complex calculations.

— **Quartiles**: Quartiles serve as dispersion measures for skewed/non-normally distributed data. The median value describes the value below which 50% of the data points lie. Below the 1st quartile lies 25% and below the 3rd quartile 75% of the data. The *interquartile range* contains all the values between the 1st and 3rd quartiles, or the middle 50% of the data. Quartiles of data that are not normally distributed are typically displayed in so-called boxplots. The box represents the interquartile range, while the data points below the 1st quartile (0–25%) and above the 3rd quartile (75–100%) can be displayed in the "whiskers" and allow the range of the entire data (0–100%) to be visualized (◘ Fig. 7.2.). Outliers are typically plotted as points outside this range rather than extending the whiskers to include them.

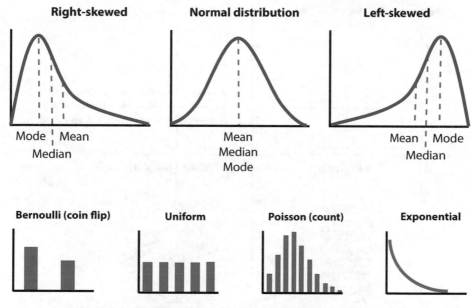

□ Fig. 7.1 Selected data distributions. The bell-shaped symmetric distribution, shown in the upper middle, corresponds to a normal distribution. Depending on the data distribution the values for mean, median and mode can vary substantially. Other common data distributions include e.g. the Bernoulli, Uniform, Poisson, and Exponential distribution

7.1.2 **Inferential Statistics**

It is important to plan how to analyse the data already when establishing your prediction(s), selecting the behaviours to be observed and recorded, and the overall study design (▶ Chaps. 2 and 3). After the data collection, it may be necessary to reassess if the initial analysis plan is still appropriate, for example by evaluating if the data distribution is as expected (□ Fig. 7.1). In order to subsequently test a prediction, inferential statistics must be used, for example to evaluate whether the means of two treatment groups differ 'significantly'. Indeed, inferential statistics play a key role to support scientific statements that are derived from numerical data, as it is common in behavioural biology. But what do we mean when we say 'significant'? **Statistical significance commonly is expressed with the p-value, or the probability that the data are consistent with the null hypothesis** (i.e. that there is no effect, ▶ Chap. 2). In other words, it measures how likely it is that any observed effect in the data is due to chance rather than a true difference in the population. Importantly, p-values can never prove or disprove hypotheses, they are simply a tool for assessing the strength of the evidence against a hypothesis. For example, a p-value of 0.04 indicates a 4% probability that the effect you observed in your sample is a result of chance and that, in fact, there is no effect in the overall population. A statistically-based rejection of the null hypothesis may

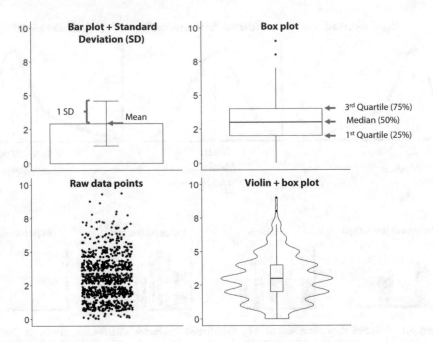

□ Fig. 7.2   Four different ways of visualising the same data (1000 datapoints drawn from a Poisson distribution). Starting top-left and going clockwise there is a bar graph with error bars based on the mean and standard deviation of the data. Next, there is a boxplot, which does a much better job of visualising the data distribution, especially for data that do not follow a normal distribution. Boxplots typically show the median and the 1st and 3rd quartiles. The whiskers indicate the entire range from 0 to 100% of the data, or (as in the example) the range excluding any outliers, which are then plotted as points. The third graph visualises the raw datapoints. This is the most detailed view of the data but can become overwhelming with large datasets. Finally, violin plots are similar to box plots but also indicate the kernel probability density at different values. When you turn this graph 90° you will see it resembles the Poisson data distribution in □ Fig. 7.1. Note that it can be useful to combine different types of graphs, for example a box plot combined with a violin plot, as shown in the last graph

thereby indirectly support your prediction. It is common to reject the null-hypothesis when the p-value is below 5% (p = 0.05), but more stringent thresholds (e.g. 1% or 0.1%) or other less 'arbitrary' approaches are increasingly common (e.g. Amrhein et al. 2019).

There are many different types of statistical tests that will test if and how significantly different or correlated the data are. To get started, it is helpful to keep your original hypothesis and predictions in mind. This is because a good prediction will indicate the dependent variable (often on the y-axis) and the independent variable (on the x-axis). For example, you may predict that birds visit feeders more when the temperature is lower. In this case the number of feeder visits is the dependent variable and temperature the independent variable. The dependent and independent variables will be the core-input for your statistical test. Furthermore, your prediction can also guide you to which type of statistical test to choose. Many predictions in animal behaviour fall into one of following categories:

1. **Do treatment groups differ in behaviour?** Prediction: Cats with tasty food eat more than cats with less tasty food. Dependent variable: Quantity of food eaten. Independent variable: Food type.
2. **Does a behaviour correlate with another variable?** Prediction: Shyer guppies solve problems faster. Dependent variable: Problem-solving speed. Independent variable: Shyness.
3. **Does a certain behaviour or choice occur more than expected?** Prediction: Chicks pass a novel object more often from the left than from the right. Dependent variable: Choice of direction. Random expectation: 50:50.

To find statistical support for your predictions, appropriate statistical tests must be applied (◘ Table 7.1). The predictions listed above correspond to the following three types of often used statistical tests:
1. Group comparisons
2. Correlations
3. Probability tests

Which statistical test to use depends not only on the question but also on the data. Some tests (e.g. t-tests) are "parametric", which assume that these data are approximately normally distributed (i.e. data have a frequency distribution in the form of a bell curve, ◘ Fig. 7.1). Parametric tests are based on the actual data values and can detect more subtle effects but can become unreliable when applied to (very) non-normal data. For non-normal data, non-parametric tests, which are based on ranked-data, should be used instead (◘ Table 7.1): common example for simple comparisons include the Mann-Whitney U test and Spearman correlation test. In non-parametric tests, the original data are converted to rank data. That is,

◘ **Table 7.1** Parametric and non-parametric statistical tests

Situation	Normal data (parametric)	Non-normal data
2 treatments/groups, same individuals	Paired t-test	Wilcoxon matched pairs test
2 treatments/groups, different individuals	Two-sample t-test	Mann-Whitney test
>2 treatments/groups, same individuals	ANOVA (one-way, repeated measures)	Friedman's test
>2 treatments/groups, different individuals	ANOVA (one-way, independent measures)	Kruskal-Wallis test
One group against an expected value	One-sample t-test	One-sample Wilcoxon rank sum test
Correlation between variables	Pearson	Spearman
Probability of count distribution across categories	Not Available (only categorical or ordinal data)	Chi-square test, Fisher exact test, Binomial test

the data is ordered by size, and then ranks are assigned to the data. The number series 2, 3, 4, 5, 17, 95 would thus be converted into the rank series 1, 2, 3, 4, 5, 6. This normalizes absolute differences between adjacent values (e.g. the difference between 2 and 3 is 1 and between 17 and the next value of 95, then is also 1). As a result, substantial information content of the original data gets lost, so that if possible the data are analysed with the original values. One can check if the data deviates from a normal distribution using visual inspection of the data distribution, for example through a histogram or q-q plot, and by conducting a normality test (e.g. Shapiro-Wilk test).

Group comparisons
When treatment groups have different individuals, an independent two-group comparison such as a t-test (or a Mann-Whitney U test) is appropriate. When we have more than two groups to compare we can conduct an Analysis of Variance "ANOVA" (or for non-parametric data a Kruskal-Wallis test). When each individual received both treatments, a paired two-group comparison, such a Paired t-test (or for non-parametric data a Wilcoxon test) is suitable. A paired test corrects for potential individual differences in behaviour (some cats always eat a lot) and so provides more power to detect differences between treatments. The best way to visualise group comparison data is via box-plots (avoid bar plots, Weissgerber et al. 2015).

Correlations
We can test if a certain behaviour (e.g. problem-solving speed) relates to another continuous variable (e.g. shyness) using the Pearson correlation test (or for non-parametric data the Spearman correlation test). The best way to show correlational data is with a scatter plot and a trend line.

Probability of observed count distributions
A simple example of such an analysis would be testing if chicks pass an object from the left significantly more often than by chance. If the chicks were behaving randomly, you would expect 50% of them to pass on the right and 50% to pass on the left. In this case we could use a one-sample Chi-square (χ^2) test to check whether any deviation from 50%, for example 140 out of 200 chicks (70%) passing the object from the right, is large enough to conclude that chicks have a side preference.

A more complex case could be investigating whether the distribution of a certain behaviour differs between groups. For example, we might want to know if male chicks are more likely to go right in a behavioural test than female chicks. First, consider what would happen if males and females do not differ in their side preference. In this case we would expect the proportions of male and female chicks passing the object on the right to be the same. For instance, if we found in a previous test, when not accounting for sex, that 70% of all the chicks go right, our 'null' expectation would be that the percentage of chicks passing right will be 70% in both male and female. Yet, when analysing the data, we find that 90 out of 100 male chicks (90%) pass the object from right while only 50 out of 100 females

◘ **Table 7.2** Contingency table for side bias in male and female chicks. Expected counts in brackets

	LEFT	RIGHT	*TOTAL*
MALE	**90** (70)	**10** (30)	*100*
FEMALE	**50** (70)	**50** (30)	*100*
TOTAL	*140*	*60*	*200*

(50%) pass it from the right. We can test if this difference between males and females is significant by conducting a Chi-square (χ^2) test with the raw count data. This test compares the observed counts to the expected counts. The best way to show such data is with bar charts showing proportions or percentages on the y axis and a so-called contingency table (◘ Table 7.2).

Box 7.3. The statistical software R
The free open-source software program R (R Core Team 2023) is commonly used by behavioural researchers for data analysis (► http://r-project.org). R offers a wide variety of packages and functions for all kinds of data management, data exploration, and data analysis, ranging from simple tests to complex models. There is also a large R user online community that provides answers to questions you may have about the program or specific functions and there are several free introductory tutorials available. Moreover, through R, and especially the package *ggplot2* (Wickham 2016), you can create detailed graphs in almost any shape or form to illustrate the results.

7.2 Presentation of Scientific Results

A study is only noticed by other scientists and the public when it is presented or published. Giving a presentation during your studies should be seen as a valuable opportunity to practice the presentation of scientific results. There are essentially three forms of presenting studies: posters, oral talks, and written publications. Talks or posters are often the first step in the presentation of scientific results, which is usually followed by writing the study up as a publication in a peer reviewed scientific journal. All three forms have their function and are briefly summarised below.

7.2.1 Oral presentations

In addition to the actual transfer of knowledge, the aim of a talk is to draw the audience's attention to the project. It is important to **be enthusiastic about the content and to be sufficiently familiar with the subject matter to be able to answer**

broader questions. The listeners' attention can be increased by regular eye contact between the speaker and the audience. Standing with your back to the audience or keeping your eyes on the screen does not make a good impression and does not promote the attention of the audience. Keyword-like notes can be helpful for a talk, but it is usually more comfortable for the presenter and for the audience if the presentation is given freely.

To make it easy for the audience to get into the subject matter, it is important to take the time to explain the theoretical and practical context of the topic. A general theoretical introduction is key to make it clear to the listener what contribution the study makes to the bigger picture. If the talk is limited to the specific study without explaining the general context it makes it difficult for non-specialists to get engaged or understand the study. Therefore, it is best to start conceptually broad and then progressively narrow the scope of the introductory information to the details specific to the study, naturally culminating in the presentation of the study's research questions or hypotheses and predictions. This structure creates a solid foundation of the topic for the listeners and enables them to interpret the results that will be subsequently revealed. When presenting the methods, only those details that are necessary for understanding the study need to be mentioned, for example the experimental design and the sample sizes. Above all, it is important to justify why something was done. Specific detailed questions on the methods can often be better clarified during questions after a talk. **In terms of content, a talk should always be tailored to the audience**, i.e. you can assume basic knowledge of your topic for a specialist audience, but explain the basics in the introduction when presenting for a less specialist audience (▶ Box 7.4).

Box 7.4: Important Points for Designing Appealing and Informative Presentation Slides
- **Uniform design** of the slides (e.g. font and colours).
- Choose **colour-blind friendly** colours that are rich in contrast and easy to recognise. Avoid 'screaming' or neon colours. **Avoid distraction**.
- Keep the lay-out **simple but attractive**.
- Present content briefly and concisely in keywords. **Avoid too much text**.
- Choose fonts (type and size) that are **easy to read**, also in graph labels.
- Design graphs clearly so that **the key result can be easily grasped**. Ideally graphs also immediately inform about the data distribution (e.g. by using boxplots) and study design (e.g. by connecting paired data points by lines). Axis-labels' font size should be sufficient large so that the audience can read it.
- If you use tables, **keep them simple**. The audience should be able to grasp the contents very quickly. Large and complex tables should thus be avoided.
- Minimize or avoid the use of playful animations. **Animations should have a purpose** and not distract or take too much time. More often than not they can be left out.

Pictures say more than words. Illustrations can therefore be a powerful way of simply conveying important information. However, for the use of illustrations and graphs to be effective, they must be clearly readable (e.g. sufficiently large axis labels) and should be carefully explained. Start with explaining what is on each of the axes. What do the datapoints represent (e.g. groups, individuals, or observations)? And what is the key result the audience should take away from the graph? Even for the explanation of a simple graph, allow sufficient time for the listener to process the information shown. Help the audience by using a pointer and guide them through the graph. Above all, focus on the graphs that help answer the research question(s) posed at the start of the presentation. Be selective in what you show, as there is only so much the audience can take in. After presenting the main results, discuss what the results mean for the bigger picture along with a (brief!) critical discussion of potential improvements for the study. Finally, talks often finish with a summary slide followed by an acknowledgements slide. However, a summary slide can be a helpful aide during the questions round, therefore it can be useful to go back to the **summary slide during the questions to stimulate discussion.** Alternatively, you can start, rather than end, the presentation by thanking the people who made it possible. Presentations are often made using Powerpoint from Microsoft Office©, but there are free alternatives such as Impress from LibreOffice.

7.2.2 Making Scientific Posters

The aim of a poster is to draw attention to your study and its take-home message. Especially at larger conferences, less is more. Here, the first aim is to attract people to your poster and the second to quickly convey the main conclusion. Since you will be standing next to your poster at some point, you can provide poster visitors with any (methodological) details they may wish to know about your study. Even more than during presentations, it is key to focus on only presenting the main results and the main message. Posters with a lot of text are often avoided because it takes the audience too long to understand them. Also keep the (scientific) background of your audience in mind. Are they experts in your area of research or do you need to provide them some additional background to your study? In other words, what does the audience need to understand the value of the study and how can it be presented in a way that the audience can grasp it quickly? It is important that the font size is large enough to be easily readable, even from a distance. Still, provide sufficient details of your study so that readers can understand it, even when you are not standing next to it. You can also provide A4 hand-outs of your (extended) poster or add QR codes/links to more information (e.g. your published study or a pdf of the poster). The design of posters is often a personal matter and, in contrast to scientific texts, there are no formal guidelines. Overall, a poster should convey a scientific message aided by an attractive design.

Here are some tips to take into consideration. Include a clear, short summary of one or two sentences. You may even start with the main conclusion or take-home message to peak the audience's interest. The introduction should be short, only a few lines, and lead to a clear research question or hypothesis. Reduce the methods section to the absolute essential. A helpful way of doing so is by using illustrations of the experimental setup and/or study design. Focus your results, by only presenting the key finding related to your research question and conclusion, accompanied by simple and informative graphs and the key statistics. Keep the discussion of the findings short and clear, again only a few lines max, but be careful to not overstate the conclusions.

7.2.3 Writing Scientific Texts

At the end of each scientific study, **researchers usually present their work as a scientific article in a peer-reviewed journal.** Scientific publications are the decisive endpoint of a study. The manuscript submitted to a peer-reviewed journal will be reviewed by specialists in the field and, if necessary, the researchers revise it according to the questions, comments and suggestions of the reviewers. Once the editor is satisfied with the changes and responses of the researchers the manuscript can be published. There are many scientific journals, and most differ in their formatting and the permitted number of words. So, before writing a manuscript, already consider which journal you aim to submit to so that you can comply with the journal-specific guidelines from the onset.

Scientific works follow formal guidelines. Most follow an Abstract-Introduction-Methods-Results-Discussion structure (▶ Box 7.5), although a few journals have the Methods at the end. A scientific manuscript, article, or report usually begins with an abstract. The **abstract**, often around 250–300 words, gives a summary of the entire work. In a few lines an abstract conveys the main background, methods, results, and the discussion. The abstract enables readers to quickly gauge if the work is of interest to them and what the most important findings are. Next follows the **introduction**. The introduction should begin with the scientific background of the project. This usually means starting with the bigger picture (e.g. beyond your study species) and incorporates relevant existing theory in animal behaviour research and important findings from previous studies. The introduction should thus include the most relevant references to existing work, primarily empirical work. Starting broad, the introduction should, in two or three paragraphs, zoom in to an important gap in knowledge. When writing the introduction, do not only make clear what is still missing but also explain why it is important that this knowledge gap (i.e. your research question) is filled. An introduction should purposefully lead to the main research question and hypothesis. After this, one commonly introduces and justifies their choice of study species and lays out the specific predictions for the study. In a few lines briefly state (in one or two sentences) how these predictions were tested. A good introduction conveys to the reader why the topic is interesting and important.

> **Box 7.5: Common Structure of a Scientific Publication**
> 1. Title
> 2. Authors with affiliations and contact details to corresponding authors, ORCID* ID's
> 3. Abstract
> 4. Keywords
> 5. Introduction
> 6. Methods
> 7. Results
> 8. Discussion
> 9. Acknowledgement
> 10. References
> 11. Declarations of authors' contributions and conflicts of interest
> 12. Ethical Note, for the work in animal experiments (sometimes also part of the methods)
> 13. Supplementary data (e.g. further details, raw data)
>
> * ORCID: Open Researcher and Contributor ID (▶ www.orcid.org)

The **methods section** describes what was done to answer the research question, **describes** how it was done, **justifies** the approach, and explains how the data were analysed. The two most important aims of a **methods** section are to allow (1) evaluation of the objectivity, reliability and validity of your approach, and (2) replication of your study. Include enough detail so that colleagues in your field could repeat your study, but avoid details that are not relevant to the study outcome. The methods section is written in past tense and the methods of a behavioural study often include sub-sections, such as a description of the study site and species for field studies, and of the animals and housing conditions for lab studies. Be sure to include the sample size and if/why certain animals were excluded. Then follows the broader study design and the specific experimental, observation and/or measurement procedures (including references). If relevant, it can be helpful to include schematic sketches of the test setup or the study design. Next, the detailed statistical analysis, which may include its own subsections (e.g. the tests used per research question and how test assumptions were evaluated), is presented. Do not forget to mention the statistical software and version number. In addition, the methods section will include information about (ethical) permits, what was done to maximize the welfare of the animals, and a statement about the data availability.

The outcome of the study is described in the **results section**. Focus here on the results that are most relevant to answering the research question(s). Any other findings that are only supportive or less relevant can go to the supplementary material/appendix. When writing the results, try to keep the animals the subject of your sentences. For example, write "male cats ate significantly faster (M = mean, SD = standard deviation) than female cats (M = mean, SD = standard deviation) during treatment A (t-test, t(degrees of freedom) = [t-value], p = [p-value], n = [sample size]; Fig. 2)", rather than "the t-test showed a significant effect of sex in treatment A (t-test, t(df) = [...], p = [...], n = [...]; Fig. 2)". An active writing style with the animals as the subject makes it easier for readers to understand the results. Figures, with captions below, and tables, with captions above, help to illustrate and summarise the data. Tables and figures should be formatted and captioned in such a way that they can be understood without knowing anything else of the study. Avoid redundancy, so do not present the same statistics in both the text and a table or in both a figure and a table. Refer to specific tables and figures between brackets after a relevant results statement. Do not make figures the topic of a sentence.

In the **discussion,** the results are critically examined and interpreted in the light of the existing literature. While in the introduction you start broad and then zoom in, in the discussion you do the opposite. The discussion starts with a summary of the most important results related to the research question(s) and a statement of whether the results support or reject the initial hypothesis. Next, the main results are discussed/interpreted one by one and it is made clear how the findings contribute to the scientific field. You can offer alternative explanations in case the hypothesis was rejected. Furthermore, it is important to discuss which aspects are still unclear and which new questions arise from the findings. A critical reflection of potential problems with a study can best be integrated throughout the discussion rather than summarized in a paragraph of its own.

The **references** section lists all the works that you have cited in the text. The formatting of the references must be uniform, but the specific citation style differs between the journals. An often used format is the so-called APA style:

Last name, First Initial. (Year). Article title. Journal Title, Volume number(Issue number), First and last page number of the entire article. Doi (Digital Object Identifier).

In the **acknowledgements**, thanks are given to anyone who helped to carry out the study, but whose contribution was not sufficient to be named as co-authors. If relevant, thanks should also be given to funders of the study.

The effort required for writing scientific articles should not be underestimated. In order to get into writing, it is a good idea to start with 'easy' parts such as the methods and results sections. In addition, it helps to make a structure of one topic sentence/keyword per planned paragraph for the introduction and the discussion. For some writers it can also help to start with the abstract, so that the central theme of the article is clear in your mind when writing the other sections.

7.3 Literature Search

In order to plan a study, it is essential to know the existing literature on the subject. Literature search is time-consuming. Current literature is available online via search engines and online data bases (e.g. Web of Science, Scopus, PubMed, Google Scholar) as well as via library catalogues of university/institute libraries. Many online databases offer means to conduct simple or advanced literature searches, allowing you to search for certain combinations of keywords, alternatives for keywords, or exclude keywords. Start simple and become more specific to reduce the number of hits to a feasible number to scan. The reference lists of relevant publications are also often helpful. Currently there are also websites (e.g. connectedpapers.com) that can give you suggestions for related work based on a relevant paper you provide as input. Scanning which publications are relevant in relation to your own work is often the most time-consuming component of a literature search. It is an acquired skill to be able to quickly review the relevance of a paper (a good abstract helps!) and assess its quality. Be critical and review if the authors can truly draw their conclusions based on their set-up, analysis, and results. Peer-review should have filtered out most bad quality studies but is not foolproof. Also, some journals generally have better quality control than others. The journals 'Animal Behaviour' and 'Behavioral Ecology' are traditionally among the leading journals in their field, but there are many other, broader, journals as well as more specific ones. Since journals differ in quality, reputation, the strictness of review process, as well as in the breadth of the readership it is important to take the time to consider which journal is the most suited one to submit a given study.

References

Amrhein V, Greenland S, McShane B (2019) Retire statistical significance. Nature 567:305–307

Harris M, Taylor G, Taylor J (2007) Startwissen Mathematik und Statistik: ein Crash-Kurs für Studierende der Biowissenschaften und Medizin. Springer spectrum, Heidelberg

Jones E, Harden S, Crawley MJ (2022) The R book. 3rd edn. Wiley, Hoboken, NJ

Quinn GP, Keough MJ (2002) Experimental design and data analysis for biologists. Cambridge University Press, Cambridge

R Core Team (2023) R: a language and environment for statistical computing. R Foundation for Statistical Computing, Vienna, Austria. ▶ https://www.R-project.org/

Wickham H (2016) ggplot2: elegant graphics for data analysis. Springer, New York

Weissgerber TL, Milic NM, Winham SJ, Garovic VD (2015) Beyond bar and line graphs: time for a new data presentation paradigm. PLoS Biol 13:e1002128

Zar JH (2013) Biostatistical analysis: pearson new international edition. Pearson Higher Ed., Upper Saddle River, NJ

Supplementary Information

© The Author(s) and Friedrich-Loeffler-Institut, under exclusive license to Springer-Verlag
GmbH, DE, part of Springer Nature 2023
M. Naguib et al., *Methods in Animal Behaviour*,
https://doi.org/10.1007/978-3-662-67792-6

Index

Printed in the United States
by Baker & Taylor Publisher Services